Springer Series in Applied Biology

Food Freezing: Today and Tomorrow

Springer Series in Applied Biology

Series Editor: Prof. Anthony W. Robards PhD, DSc, FIBiol

Proposed future titles:

Biodegradation of Natural and Synthetic Materials
Ed. W. B. Betts

Separation and Immobilisation of Biomolecules
Ed. U. B. Sleytr

Gene Transfer in Eukaryotic Cells
Ed. J. R. Warr

Critical Loads of Environmental Pollutants
Ed. M. J. Chadwick

Food Freezing:
Today and Tomorrow

Edited by W. B. Bald

Springer-Verlag
London Berlin Heidelberg New York
Paris Tokyo Hong Kong

W B Bald
Wolfson Unit for Applied Cryobiology, Institute for Applied Biology, Department of Biology, University of York YO1 5DD

Series Editor
Professor Anthony William Robards, BSc, PhD, DSc, DipRMS, FIBiol
Director, Institute for Applied Biology, Department of Biology, University of York, York YO1 5DD, UK

Cover Illustration: Micrograph of ice-cream showing ice crystals (cubic blocks) surrounding cellular air pockets - courtesy of Dr Ashley Wilson, CCTR, Institute for Applied Biology, University of York

ISBN 3-540-19668-4 Springer-Verlag Berlin Heidelberg NewYork
ISBN 0-387-19668-4 Springer-Verlag New York Berlin Heidelberg

British Library Cataloguing in Publication Data
Food freezing
1. Frozen food
1. Bald, William Balfour, 1933- Apr. 4 -
664.02853
ISBN 3-540-19668-4

Library of Congress Cataloging-in-Publication Data
Food freezing: today and tomorrow / edited by W. B. Bald
p. cm.- (Springer series in applied biology)
ISBN 0-387-19668-4
1. Frozen foods. 1. Bald, W. B. II. Series.
TP372.3.F66 1989
664'.02853 - dc20 91-4522
CIP

Printed in Great Britain

Set by Institute for Applied Biology, Department of Biology, University of York
Printed by The Alden Press Ltd., Osney Mead, Oxford

Foreword from Series Editor

The Institute for Applied Biology was established by the Department of Biology at the University of York to consolidate and expand its existing activities in the field of applied biology. The Department of Biology at York contains a number of individual centres and groups specialising in particular areas of applied research which are associated with the Institute in providing a comprehensive facility for applied biology. Springer-Verlag has a long and successful history of publishing in the biological sciences. The combination of these two forces leads to the "Springer Series in Applied Biology". The choice of subjects for seminars is made by our own editorial board and external sources who have identified the need for a particular topic to be addressed.

The first volume, *"Foams: Physics, Chemistry and Structure"*, has been quickly followed by *"The 4-Quinolones: Antibacterial Agents in Vitro"* and now *"Food Freezing: Today and Tomorrow"* and the range of seminars will extend from genetic engineering to ecotoxicology and biodegradation to plant micropropagation. The aim is to keep abreast of topics that have a special applied, and contemporary, interest. The current volume, dealing with Food Freezing, addresses many of the crucial problems in an industry where concerns with improving quality, health and safety, and increasing legislative controls, all need to be fully considered.

Up to four volumes are published each year through the editorial office of *IFAB Communications* in York. Using modern methods of manuscript assembly, this streamlines the publication process without losing quality and, crucially, allows the books to take their place in the shops within four to five months of the actual seminar. In this way authors are able to publish their most up-to-date work without fear that it will, as so often happens, become outdated during an overlong period between submission and publication.

The applications of Biology are fundamental to the continuing welfare of all people, whether by protecting their environment or by ensuring the health of their bodies. The objective of this series is to become an important means of disseminating the most up-to-date information in this field.

York, December 1990 A. W. Robards

Editor's Preface

The following chapters represent the contributions of invited Speakers to an Advanced Study Seminar organised by the Institute for Applied Biology at the University of York during April 1990. The main objective of the Seminar was to bring together food scientists and engineers to discuss their mutual problems and to seek new directions for possible solutions.

Food freezing is very much a multi-disciplinary process which will require in the future even higher standards of quality and food texture than are acceptable at present. New legislation will demand better control of the growth and viability of micro-organisms in frozen foodstuffs.

The first few chapters of the book cover some of the fundamental aspects of frozen foods which must be fully understood before the practical freezing methods, used in the food industry, are to be improved and made more efficient.

Water in foodstuffs and its physical state is characterised by biopolymer interactions and its transport between regions of different water activity. In frozen foods a certain proportion of the water remains unfrozen and therefore the extent and distribution of this water in different products must be known if drip loss is to be minimised. This loss of water is one of the most important features of frozen foodstuffs.

Cooling inhibits microbial growth by the reduction of metabolic rate and by the progressive decrease in water activity. At the normal temperatures used for frozen storage microbial growth is inhibited but long term storage is often limited by the chemical and physical changes which take place. Microbiological aspects are therefore very important when studying the injury to microorganisms during the freezing process and subsequent storage.

The size of ice crystals which form in different products during freezing and which continue to grow during storage have a crucial effect on the quality of the final product. The crystallisation of ice from sucrose solutions and the effect of polymers on the rate of ice crystal growth are also very important factors to be considered in frozen foods.

Ice crystal size is directly related to the cooling rate applied to any product and this cooling rate can only be accurately determined if the thermophysical properties of different foodstuffs are known throughout the relevant temperature range. A chapter of the book is therefore devoted to this important aspect of the overall problem.

The structure of food products after freezing can only be fully envisaged by using either low temperature electron or optical microscopy and consequently two chapters in the book discuss these specialised techniques. The remaining chapters are concerned with special applications such as the freezing of fish, fruit and vegetables and ice cream and the engineering equipment used to achieve these specific objectives.

This advanced study seminar may not have answered all the technical problems facing the food freezing industry in the future but if it has highlighted the areas requiring further detailed research the seminar will have achieved its major objective.

Finally I would like to extend my thanks to the authors for their time and efforts involved in the preparation of these manuscripts.

W. B. Bald

Contents

Contributors

Dr. R. E. Angold
RHM Research and Engineering, The Lord Rank Research Centre, Lincoln Road, High Wycombe HP12 3QR

Mr. W. B. Bald
Wolfson Unit for Applied Cryobiology, Institute for Applied Biology, Department of Biology, University of York, York YO1 5DD

Dr. P. S. Belton
AFRC Institute of Food Research, Colney Lane, Norwich NR4 7UA

Dr. G. Blond
ENSBANA, Université de Bourgogne, Departement de Biologie Physico-Chimique, Campus Universitaire Montmuzard, LF-21000 Dijon, France

Dr. M. H. Brown
Unilever Research, Microbiology Section, Colworth Laboratory, Colworth House, Sharnbrook, Bedford MK44 1LQ

Mr. B. Colas
Ecole Nationale Supérieure de Biologie Appliquée à la Nutrition et à l'Alimentation, Université de Bourgogne, Dijon, France

Mr. L. Eek
Frigoscandia Contracting AB, Helsingborg, Sweden

Mr. D. W. Everington
APV Baker Limited (Freezer Division), Stephenson Way, Thetford, Norfolk IP24 3RP

Dr. B. W. W. Grout
Novalal plc, Merks Estate, Great Dunmow, Essex CM6 3BD

Dr. C. Holt
Unilever Research, Colworth Laboratory, Colworth House, Sharnbrook, Bedford MK44 1LQ

Mr. K. Hughes
Planer Products, Windmill Road, Sunbury on Thames, Middlesex TW16 7HD

Mr. J. Lavety
Ministry of Agriculture Fisheries and Food, Torry Research Station, P. O. Box 31, 135 Abbey Road, Aberdeen AB9 8DG

Dr. M. R. McLellan
Cell Systems Limited, Orwell House, Cowley Road, Cambridge CB4 4WY

Dr. C. A. Miles
AFRC institute of Food and Research, Langford, Bristol BS18 7DY

Mr. J. P. Miller
Air Products plc, Brunel Science Park, Kingston Lane, Uxbridge, Middlesex UB8 3PQ

Dr. G. J. Morris
Cell Systems Limited, Orwell House, Cowley Road, Cambridge CB4 4WY

Dr. G. W. Rodger
Marlow Foods, P. O. Box 1, Belasis Avenue, Billingham, Cleveland TS23 1LB

Dr. E. M. A. Willhoft
Epsom Technical Group Services, 41 Higher Green, Epsom, Surrey KT17 3BB

Dr. A. J. Wilson
CCTR, Institute for Applied Biology, Department of Biology, University of York, York YO1 5DD

Chapter 1

The Physical State of Water in Foods

P. S. Belton

Introduction

Foods are characterised by their heterogeneous nature, typically they contain many components, more than one phase and spatial heterogeneity on a variety of scales. A further complication is that almost all foods are unstable chemically and physically and are thus highly time dependent in their properties. A good example is a loaf of bread (Fig. 1.1); the major components are wheat flour and water; minor components are sodium chloride, air, sugar, and possibly alcohol from fermentation; there will also be yeast cells in various states of decomposition and damage. The yeast is a highly complex and organised system, it is not however a major component. The chemically most complex major component is flour. This contains gluten and other proteins, starch, lipids and other polysaccharides, each one of these components is itself chemically heterogeneous. The air, although a small component by weight, plays a major role in determining the size and shape of the loaf and is responsible for the foam structure of the crumb.

The chemical complexity is distributed between different physical states. The vapour phase contains air, water and flavour volatiles. The solid phases contain both crystalline and amorphous material - typically some small amount of starch will be crystalline in a fresh loaf but the amount will increase with time - the proteinaceous components are likely to be amorphous. Whilst there are no separate liquid phases there is good evidence (Belton *et al.* 1988) to suppose that some lipids associated with the gluten will be more liquid-like than solid-like in their behaviour.

On the large scale spatial heterogeneity is apparent between the outer crust, which has been subjected to the most intense heat in cooking, and the less heated internal crumb. There will also be a gradient of water concentration between crust and crumb due to evaporation during cooking. On the scale of millimetres air bubbles are trapped in the foam structure. This structure results in high surface areas of the solid phase and

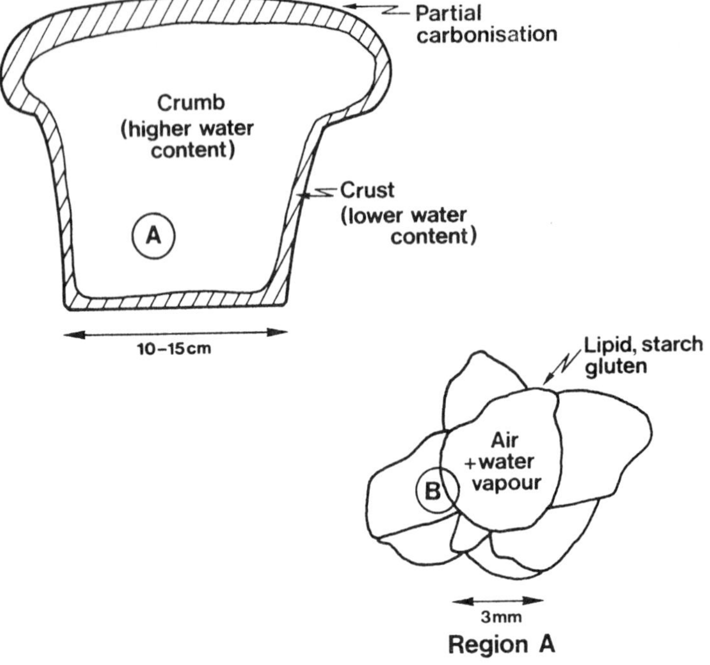

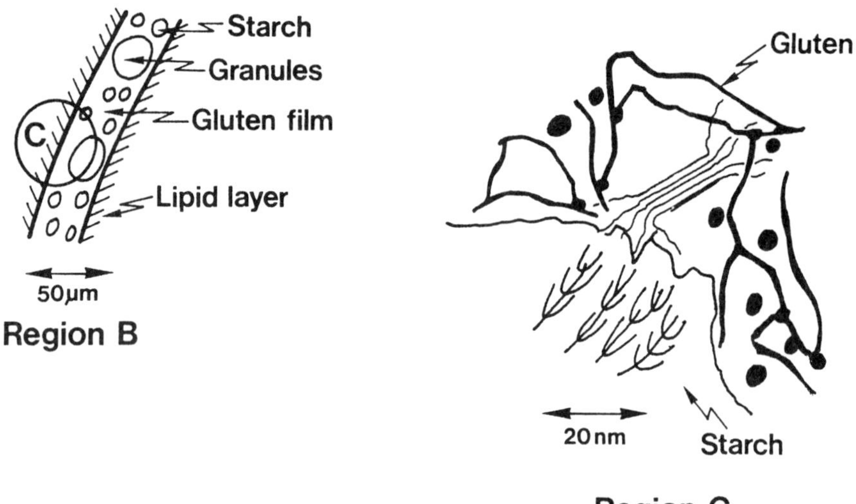

Fig. 1.1. The heterogeneity of food illustrated by a slice of bread. The scales are indicated in the diagram.

typifies another characteristic of foods - their interfacial nature. The solid phase contains starch rich and protein rich regions on the scale of tens of microns. Within the starch region crystalline and amorphous sub regions exist on a variety of scales. Within the protein region there are a large number of different proteins and motional heterogeneity on the scale of tens of nanometres is present (Belton *et al.* 1988).

From the foregoing description it is clear that no simple statement as to the physical state of water in food is possible. The state of the water will be highly dependent on its spatial location and will vary with time. It is possible however to list the features which will have to be considered in order to obtain a general description of the water. These are:-

1. The interactions of water and biopolymers.
2. The effects of interfaces.
3. Mechanisms of water transport.

Much the same considerations will apply in frozen food systems but here the added complexity of another phase of water is present. In the next sections the factors governing water behaviour in food are considered.

Water Biopolymer Interactions

The specific details of the interactions of water with biopolymers will depend on the specific details of the biopolymers, however, in general the water in all types of systems may be divided into three main classes. These are (Berendsen 1981, Belton 1984): water that is structurally integrated with the biopolymer (Type 1 water); water that has its motion affected by interaction with the biopolymer (Type 2 water); water remote from the biopolymer and not affected by its presence (Type 3 water).

On a long timescale all these types of water will exchange with each other, however in general exchange between type 1 water and the other two may be expected to be slow whereas it is likely that exchange between types 2 and 3 water will be rapid.

When the water content is high all forms of water are likely to be present. Type 1 water will typically be associated with native proteins and polysaccharides in ordered conformations, for example helical amylopectin in crystallites. The amounts of type 1 water will depend on the details of molecular structure and may range from one or two per polymer molecule up to levels where there is a water molecule in every helical turn (Berendsen 1981). Generally it is to be expected that when the water level is reduced exchange with external water molecules, and rotational motion, will be slow. As the water content increases motion and exchange rates are likely to increase.

Type 2 water will be more generally associated with the hydrophillic groups of biopolymers. The nature and population of type 2 water has been the subject of considerable debate in the past much of which has centred around the interpretation of Nuclear Magnetic Resonance (NMR) relaxation time measurements and the nature of "bound" water. For systems where the water content is high a general concensus of interpretation has now been reached which suggests (Belton 1984, Belton 1990, Kakalis and Baianu 1988) that water affected by interaction with biopolymers undergoes anisotropic motion, with one rotational rate of the same order of magnitude as bulk water and one rate closer to that of the reorientational correlation time of the biopolymer. Thus although the motional state of the water is affected the idea that such water is immobilised is incorrect and the term "bound" is something of a misnomer.

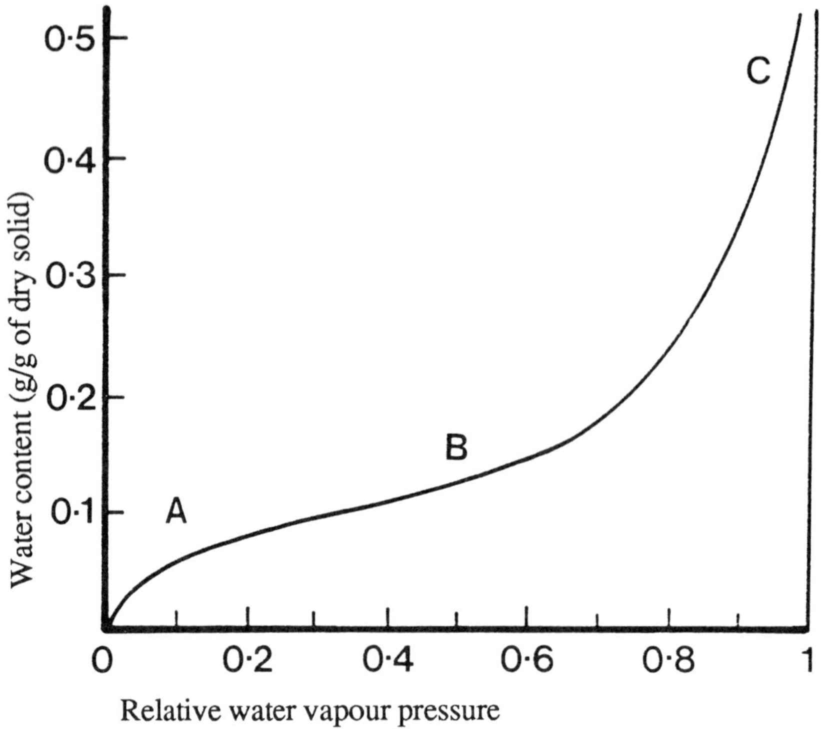

Fig. 1.2. A generalised water absorption isotherm for a foodstuff or a protein.

The amount of water involved is the equivalent of one or two monolayers of water around the biopolymer. This is of the order of 0.2 g to 0.5 g of water per g of biopolymer and is of a similar quantity to the amounts of non-freezing water observed in biopolymer systems.

When the material is dry and the water vapour pressure is increased water absorption takes place. It is useful to consider the different regions of the water absorption isotherm. A very good account has been given by van den Berg and Bruin (1981) which is followed closely here. At low relative water vapour pressures (region A Fig. 1.2) water is absorbed only onto sites immediately accessible to the vapour, and with a high affinity for water, typically polar sites. At these low water contents there is very little water mobility (Lechert 1981) and the biopolymer is not plasticised by the presence of the water. The behaviour at this stage is therefore analogous to the absorption of an inert gas on an impermeable surface. As vapour pressure increases more water is absorbed (region B Fig.1.2) water enters the system and begins to plasticise the biopolymer. This allows rearrangements to take place. In the case of starch and related materials (Tanner *et al.* 1987) crystallisation may occur. This process is inhibited at very low water content by the lack of mobility and by a lack of sufficient water molecules to provide the required water of crystallisation. A concept which is very important in explaining behaviour at low water contents is that of the glassy state.

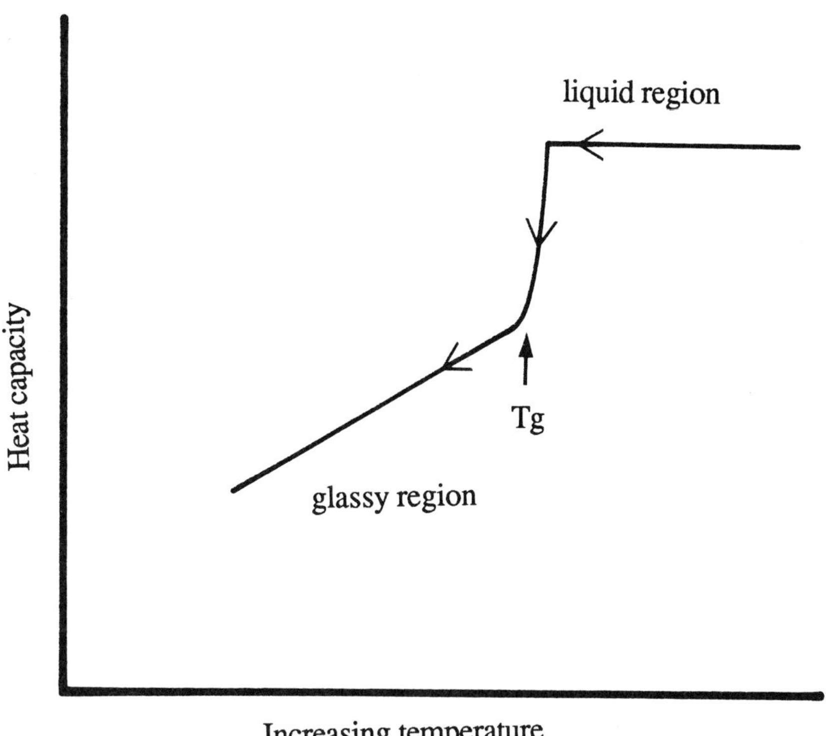

Fig. 1.3. The effect of the glass transition on heat capacity.

This concept is well known in physics (Johari 1982) but has only more recently been applied to food science by Slade and Levine (for a recent review see Levine and Slade 1989). A glass is a disordered material in which molecular motions are very slow and in which heat capacity, compressibility and expansivity are close to those in the crystalline rather than the liquid state. As a consequence of this translational motion is very slow and viscosity is very high. A typical example of a glass is a boiled sweet. Associated with glass formation is the glass transition temperature Tg (Johari 1982, Levine and Slade 1989). In order to appreciate this parameter it is best to consider the cooling of some low water content but mobile material from a high temperature where motion is still present. Examples would be a sugar syrup or an extruded product just as it leaves the dye of the extruder. If the temperature is dropped then at some point there will be a drop in the heat capacity from the liquid value to one close to the crystalline value (Fig. 1.3). The temperature at which this occurs is the glass transition temperature. Glasses are metastable with respect to the crystalline state but in general are kinetically very stable. In biopolymer systems the glass transition temperature is dependent upon water content and at a fixed temperature, may move away from the glassy state with the addition of more water. In essence the glassy state is one in which motion is inhibited by the close packing of the molecules. The difference between it and a crystalline material is that there is no long range order. Clearly glassy states are likely to be important in low temperature systems as well as low water content systems.

Region C in Figure 1.2 marks the end of the region where more specific interactions occur and the beginning of the region where void spaces begin to fill up. In this region non-specific surface effects begin to be important and these form the subject of the next section.

Interfacial Effects

The interfacial nature of foodstuffs is a major feature of their physical chemistry from the point of view of water the most important interfacial property is capillarity. This can arise in a variety of ways: it may be inherent in the food structure; it may result from mechanical fracture during cooking or handling or it may be introduced by processing or ingredient choice. Meat provides a very good example of the first type of capillarity, muscle (Wilkie 1970) contains large numbers of fibres of the order of 100 μm in diameter these in turn contain fibrils of about 1 μm diameter and the fibrils contain myofibrils of about 50 nm diameter. Within the myofibrils are parallel protein filaments in which interprotein distances are on the nanometer scale.

Mechanical fracture and damage can occur on a wide variety of scales from the macroscopic to close to the molecular level. It can range from large fissures through to microscopic cracks in the surfaces of starch granules or small regions of cell damage. Capillarity can be also introduced by processing, for example spray drying can result in fissures and cavities on a variety of scales. Alternatively ingredients may be added which result in additional capillarity in the system. Gelling agents are representative of this class of materials. Although gelling agents are typically present in low percentage amounts they may be regarded as molecularly dispersed. Thus a gel containing 1% of a polymer whose chain diameter is 3 nm will, if the chain is fully extended, be only 50 nm from its nearest neighbours.

Capillarity can affect water behaviour in three distinct ways. These are: viscous flow; capillary rise effects and vapour pressure effects. Viscous flow from a capillary is governed by the Poiseuille equation which may be written as (Glasstone 1960)

$$V_t = \frac{\pi P r^4}{8 l \eta} \tag{1}$$

where V_t is the volume flowing out per second from a cylindrical tube of radius, r, and length, l, under a pressure P. η is the viscosity of the liquid. In order to compare the effects of different radii the pressure is assumed to result purely from the weight of the vertical column of water and the length, l, is assumed to be 2r. The total volume of the capillary is then $2\pi r^3$. The quantity lost per second can then be expressed as a fraction of the total volume and

$$V'_t = \frac{V_t}{2\pi r^3} = \frac{\rho g r}{16\eta} \tag{2}$$

where ρ is the density of the fluid and the substitution $P = 2\rho g r$ has been made, g is the acceleration due to gravity. Equation 2 allows a measure of the expected loss from capillaries of different radii. The assumption $l = 2r$ takes account of the fact that small radius capillaries are likely to have less extension than large radius capillaries. Results of the calculation for a variety of radii are given in Table 1.1. Consideration of the Table shows that the time scale of drainage in small capillaries can be long but this does not explain the behaviour of many foods which exhibit water holding capacities over much longer time scales.

Table 1.1. The effects of capillarity on drainage, suction pressure and relative vapour pressure

Capillary radius(cm)	Drainage rate[a](s^{-1})	Suction pressure(Atm)	P/P_o
0.1	6.1×10^2	1.42×10^{-3}	1.000
1×10^{-4}	0.61	1.42	0.990
1×10^{-5}	0.061	14.20	0.989
5×10^{-6}	0.031	27.00	0.979
1×10^{-6}	6.1×10^{-3}	142.00	0.900

[a] see Equation 2 in the text for a definition

Implicit in the calculation is the assumption that surface tension plays no part in the flow behaviour. This is true when there is a reservoir of liquid water at each end of the capillary, however when one end of the capillary is in contact with vapour, surface effects have a profound effect on local pressure. These effects cause the phenomenon of capillary rise (Hunter 1987) and capillary suction pressure. Capillary rise is the result of the free energy gain of the liquid interacting with the wall compensating for the free energy loss associated with raising a column of liquid against its intrinsic hydrostatic pressure. The height of the column of liquid which may be supported is given, assuming that complete wetting takes place by (Hunter 1987)

$$h = \frac{2\gamma}{r\rho g} \tag{3}$$

where h is the height of the column, g is the surface tension and r the radius of the capillary at the interface. For small capillaries the height of a water column that can be supported is considerable (Hunter 1987) for a 150 nm radius capillary the height is 100 m, even for a capillary 1 mm in radius the rise is about 14 mm. It is important to note that Equation 3 only depends on the radius at the interface, thus with irregularly shaped capillaries large volumes of liquid may be supported provided that there is an unbroken column of water to the interface.

More relevant for water holding capacity of foods is to calculate the suction pressure arising when an empty capillary is placed in contact with liquid. This is equivalent to the pressure required to cause liquid to be expelled from the capillary when it has water in it. This pressure, P, is given by (Hunter 1987)

$$P = \frac{2\gamma}{r} \tag{4}$$

Table 1.1 gives an indication of the range of pressures which are obtainable. Clearly at radii below 1 μm (10^3 nm) suction pressures are such that water becomes very difficult to expel from the capillary and thus water holding becomes very tenacious.

The final effect due to capillarity is the lowering of relative vapour pressure in capillaries, this is given by the Kelvin equation (Hunter 1987) as

$$\ln P / P_o = \frac{-2\gamma\bar{v}}{rRT} \tag{5}$$

where P is the vapour pressure above a capillary. P_o is the vapour pressure above an infinitely wide capillary, R is the gas constant and T the absolute temperature. P/P_o is thus the relative vapour pressure, often referred to in the food science literature as the water activity (a_w). Table 1.1 shows the variation in a_w for a range of capillary diameters. This effect is only likely to be significant when capillary radii are very small. Since such very small capillaries are unlikely to constitute a significant amount of the volume of the foodstuff it is unlikely that capillarity effects contribute significantly to reductions in water activity at high water contents, these are therefore more likely to arise from interactions with biopolymers and solutes.

Water Transport

Foods are not systems in equilibrium. They are likely to contain gradients of water activity within them as a result of preparation and cooking or due to drying in contact with the atmosphere. As a result water transport is occurring during the lifetime of most foodstuffs. Transport can be on a very large scale. As pointed out earlier, in the case of bread there is migration from the relatively moist crumb to the dryer crust. Transport also occurs on a more local scale, there is, for example, the well known problem in confectionery of transfer of water from relatively high water content fruit components to the dryer biscuit causing problems with the biscuit texture. The differences in water activity continue down through the whole range of distances to those close to the molecular level. Transport mechanisms will depend on the details of morphology and water activity differences. In phenomena such as drip loss in meat bulk flow processes are likely to be important and rates may be determined by pore sizes. In other systems diffusion is the main source of transport. Under these circumstances the presence of a vapour phase can increase transport rates significantly since the mechanism could involve vaporisation and diffusion in the vapour phase. This can be a very efficient transport process as diffusion in vapours is about four orders of magnitude faster than that in condensed fluids.

Water in Frozen Systems

One of the major characteristics of foodstuffs and biopolymeric systems in general is that, as they are cooled, ice formation takes place but a fraction of the water present does not undergo a phase transition and remains liquid down to very low temperatures (Derbyshire 1982, Kuntz and Kauzmann 1974). The measurement of this "non-freezing" water can be very simply carried out by NMR methods (Derbyshire 1982, Kuntz and Kauzmann 1974, Nagashima and Suzuki 1984). The basis of the technique is that the proton transverse relaxation time of ice is of the order of microseconds whilst that of non-freezing water is of the order of hundreds of microseconds to milliseconds. Experimentally, therefore, all that is required is to measure the magnetisation decay curve following excitation with a radiofrequency pulse. The intercepts of the fast and slow decays, when extrapolated to zero time, give the relative amounts of protons in the solid and liquid forms. Figure 1.4 shows the general structure of the proton relaxation curve in an ice water system. In general the measurement of the fast decaying component due to ice is not easy and is not necessary since the intercept of the slow decaying component is indicative of the total amount of

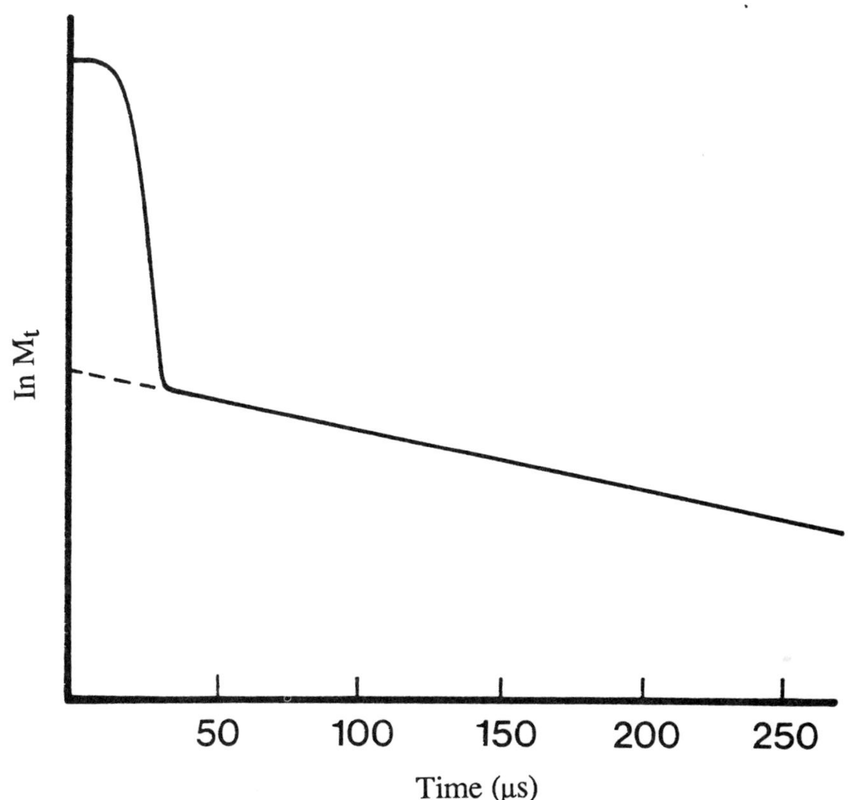

Fig. 1.4. An idealised plot of transverse magnetisation (M_t) versus time for protons in a solid liquid mixture. The solid component shows non-exponential behaviour. The extrapolation of the slow curve to zero time gives the amount of liquid present.

non-freezing water. If it is compared to the total mobile signal intensity before freezing, direct calculation of the non-freezing water content is possible. Since transverse relaxation time is inversely proportioned to the spectral line widths, very short relaxation times give very broad lines not observable in conventional high resolution NMR spectrometers, the narrow line due to mobile component is observed however. The non-freezing water may therefore be simply quantified by measurement of the intensity of the water signal remaining after freezing (Kuntz and Kauzmann 1974). However the measurement is carried out, it is important to take account of signals that might arise from mobile species other than water.

The existence of an aqueous phase after ice formation is of course normal in simple solutions. The phase persists until either crystallisation of the pure solid dissolved in the aqueous phase or until solid solutions are formed. The important point is that there is a single temperature (if pressure is kept constant) at which the liquid phase disappears. This is not the case for cooling in foods, biological systems, and biopolymer solutions. The liquid phase persists to very low temperatures and shows only a modest or zero decrease in amount as temperature is lowered. In protein solutions the amount of non-freezing water is very closely related to the amino acid composition (Kuntz and Kauzmann 1974). Indeed by measuring non-freezing water in polypeptides it is possible to predict the non-freezing water content of protein solutions with some accuracy. Typically, charged amino acids are associated with between three and seven water molecules, polar amino acids with two or three and non-polar acids with none or one. There is also one water molecule associated with each peptide link. Typically there is between 0.3 g and 0.6 g of non-freezing water per g of protein, a value which is also fairly typical of foods and biological systems (see Table 1.2).

Table 1.2. Observed and predicted hydration and non-freezing water content of protein solutions and other materials. Hydration and non-freezing water contents are expressed asgrammes of water per g of dry weight. The data are taken mainly from Kuntz and Kauzmann 1974.

Substance	Non-freezing water (NMR)	Predicted non-freezing water	Hydrodynamic hydration	Vapour pressure hydration
Lysozyme Bovine Serum	0.34	0.36	0.46	0.25
Albumin	0.40	0.45	0.43	0.32
Ovalbumin	0.33	0.37	0.15	0.30
Haemoglobin	0.42	0.42	0.50	0.37
Minced chicken[a] muscle	0.4	-	-	-
Wheat starch[b] (Pastes)	0.35	-	-	-

[a] T.E. Southon, K.J. Packer and P.S. Belton, unpublished results.
[b] P.T. Callaghan, K.W. Jolley, J. Lelievre and R.K. Wong (1983) J Colloid Interface Sci 92:332.

The question of the origins of the properties of non-freezing water has often been discussed in terms of bound water (Derbyshire 1982, Kuntz and Kauzmann 1974) or kinetic effects (Levine and Slade 1989, Franks 1982) as if they were mutually exclusive phenomena. Although the word "bound" may not be entirely appropriate, it is clear from the work of Kuntz and co-workers that the amount of non-freezing in protein solutions is predictable to a high degree of accuracy solely on the basis of amino acid content. The amount of non-freezing water is close to estimates of water of hydration by other methods (Table 1.2) which do not rely on freezing. It seems hard to believe that such agreement is coincidental and the conclusion must be that some fraction of water in protein solutions does behave differently to the bulk. This does not imply that kinetic factors are not important, Franks (1986) has pointed out that in sucrose solutions at -32°C the growth rate of ice crystals is about 0.3 μm per year thus the system is kinetically stable. However to infer, as he then does, that all non-freezing effects arise simply from the slowing of diffusion and hence ice crystal growth is a long step. It would imply that all the protein solutions listed and all the polypeptide solutions tested reached a situation where diffusion was sufficiently slowed to prevent crystallisation in a manner predictable solely on the basis of amino acid content. This seems unlikely as one would expect factors such as conformation and chain length to have an effect.

Another way of examining the effects of cooling on motion is to examine the evidence from NMR measurements. Although there are problems with the interpretation of proton NMR data (Belton 1990), to within an order of magnitude it would be expected that the NMR line width scales with the correlation time for motion and the diffusion coefficient. At 25°C proton line widths for pure water are about 0.1 Hz; at -35°C line widths are the order of 1 kHz, about 10^4 times wider. At 25°C the diffusion coefficient for water is 2×10^{-5} cm^2 s^{-1} and therefore at -35°C it will be of

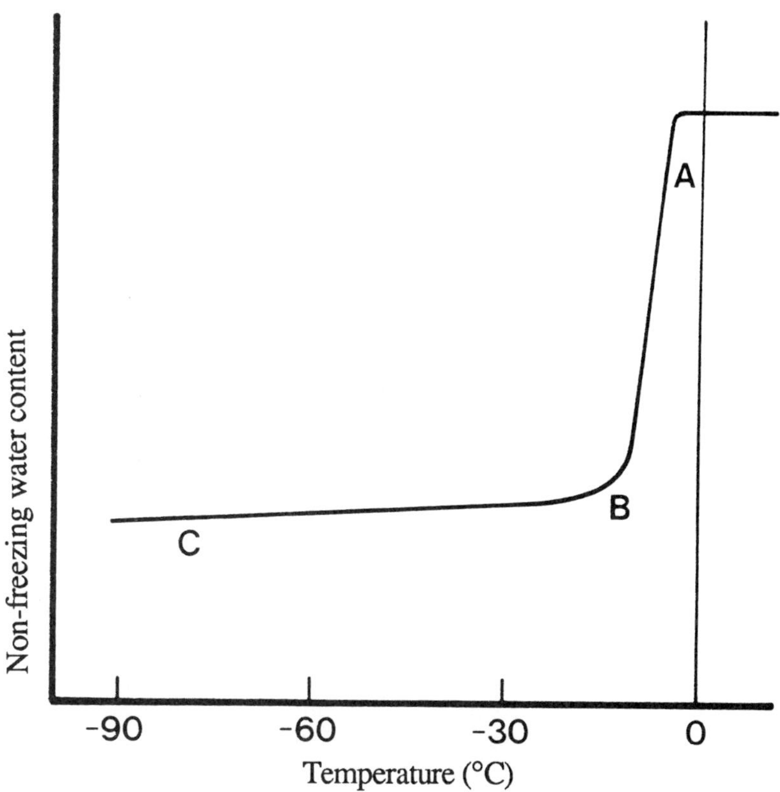

Fig. 1.5. The variation in non-freezing water content with temperature for a typical protein solution.

the order of 2 x 10^{-9} cm^2 s^{-1}, the time taken to diffuse 1 μm is therefore about 2.5 s. This means that an ice front could grow at the rate of about 1 mm per hour. This would be more than sufficient for significant changes in non-freezing water content to be observed during experiments.

A model of how protein solutions freeze may be constructed which takes account of both kinetic and hydration effects. The process is illustrated in Figure 1.5. Initially as the protein solution is cooled ice forms and the aqueous phase becomes more concentrated, the behaviour at this point is similar to that of an ordinary solution. (Region A). As temperature is lowered still further more ice is formed and the increasing concentration of the aqueous phase causes crowding of the proteins, this slows reorientational motion and inhibits crystallisation of the protein phase, water can however still diffuse to the ice interface and ice crystals grow. At some point the free energy of transfer from the aqueous to the ice phase becomes zero or very near to zero and the driving force for ice formation becomes very small. (Region B). This is the point where the vapour pressure of the aqueous phase is equal to that of the ice phase. This point may be estimated by using the data compiled by Angell (1982) for the vapour pressure of super-cooled water. When these values are compared to those of ice the ratio of water to ice vapour pressures ranges from, (necessarily) 1 at 0°C to about 0.8 at -30°C. In order for the vapour pressures to be equal the water activity for the aqueous phase must be lowered to between 1 and 0.8. At room temperature this corresponds to water contents down to about 0.2 g of water per g of protein and is likely to correspond to higher water contents at lower temperatures (Kapsalis 1987, MacKenzie 1975). This range of water contents represent a metastable state. If crystallisation were not inhibited the proteins would crystallise and a normal eutectic would be expected. Because this does not happen ice formation is similarly inhibited. As temperature is lowered still further the differences in vapour pressure of ice and water may or may not increase (Angell 1982) in any case molecular motion will be slowed down and the kinetic effects discussed by Franks will come into play. (Region C).

In food systems the motion of the larger constituents is often constrained by the structure, there is no possibility of crystallisation of the biopolymer components. The formation of ice crystals may be similarly constrained by the structure of the system, however the amounts of non-freezing water observed are similar to those observed in simple biopolymer solutions again consistent with the notion that amino acid content and the free energy of transfer from the biopolymer rich region to the ice phase plays a controlling role in determining the non-freezing water content.

Conclusions

The behaviour of water in food systems may be generally explained on the basis of straightforward and well understood physical principles. The problem is however that the current state of our knowledge does not yet permit a fully quantitative understanding of behaviour and this is what is required if it is ever to be possible to design foods which in the behaviour of water is fully predictable and controllable. Probably the main factors which frustrate the achievement of this goal are adequate experimental and theoretical methods for dealing with the complexity and heterogeneity of food materials.

Acknowledgements

This work was supported in part by the Ministry of Agriculture Fisheries and Food.

References

Angell CA (1982) Supercooled water. In:Franks F (ed) Water a comprehensive treatise Vol 7. Plenum NewYork, pp 38

Belton PS (1984) Spectroscopic methods. In:Chan HW-S (ed) Biophysical methods in food research. Blackwell, Oxford, pp 103-137

Belton PS (1990) Can NMR give useful information about the state of water in foodstuffs? Comments on Food and Agricultural Chemistry (in press)

Belton PS, Duce SL and Tatham AS (1988) Proton NMR relaxation studies of dry gluten. J Cereal Sci 7:113-22

Berendsen HJC (1981) Specific interactions of water with biopolymers. In: Franks F (ed) Water a comprehensive treatise, vol 5. Plenum, New York and London, pp 293-330

Derbyshire W (1982) The dynamics of water in heterogeneous systems with emphasis on sub-zero temperatures. In:Franks F (ed) Water a comprehensive treatise, vol 7. Plenum, New York, pp 339-430

Franks F (1982) The properties of aqueous solutions at sub-zero temperatures. In:Franks F (ed) Water a comprehensive treatise, vol 7. Plenum, New York, pp 261-270

Franks F (1986) Unfrozen water:yes; unfreezable water:hardly; bound water:certainly not. Cryo-Letters 7:207

Glasstone S (1960) Textbook of physical chemistry. Macmillan, London, pp 498

Hunter RJ (1987) Foundations of colloid science. OUP, Oxford, pp 272-277

Johari GP (1982) Glass transition and molecular mobility. In:Escaig B and G'Sell C (eds) Plastic deformation of amorphous and semi-crystalline materials. Les Editions de Physique, Les Ulis, France, pp 110-141

Kakalis L and Baianu I (1988) Oxygen 17 and deuterium NMR relaxation studies of lysozyme hydration in solution. Arch Biochem Biophys 267:829-841

Kapsalis JG (1987) Influences of hysteresis and temperature on moisture sorption isotherms. In: Rockland LB and Beuchat LR (eds) Water activity theory and applications to food. Marcel Bekker, New York, pp 173-213

Kuntz ID and Kauzmann W (1974) Hydration of proteins and polypeptides. Adv Protein Chem, 28:239-345

Lechert HT (1981) Water binding in starch:NMR studies on native and gelatinized starch. In:Rockland LB and Stewart GF (eds) Water activity:influences on food quality. Academic, New York, pp 223-245

Levine H and Slade L (1989) Interpreting the behaviour of low-moisture foods. In: Hardman TM (ed) Water and food quality. Elsevier, London, pp 71-134

MacKenzie AP (1975) The physico-chemical environment during the freezing and thawing of biological materials. In:Duckworth RB (ed) Water relations of foods. Academic, London, pp 477-503

Nagashima N and Suzuki E-I (1984) Studies of hydration by broad line pulsed NMR. Appl Spectrosc Rev 20:1-53

Tanner SF, Ring SG, Whittam MA and Belton PS (1987) High resolution solid state ^{13}C NMR of some 1-4 glucans. Int J Biol Macromol 9:219-224

van den Berg C and Bruin S (1981) Water activity and its estimation in food systems:theoretical aspects. In: Rockland LB and Stewart GF (eds) Water activity:influences on food quality. Academic, New York, pp 1-61

Wilkie DR (1970) Muscle. Edward Arnold, London, pp 3-7

Chapter 2

Microbiological Aspects of Frozen Foods

M. H. Brown

Introduction

At first sight the microbiology of frozen foods is not a very promising topic, as it is well known that frozen temperatures halt microbial growth. However, the mechanism of action and the effects of freezing and thawing on microbial death, sub-lethal injury and the recovery of microorganisms from frozen foods have been extensively investigated because of their practical importance. More recently the techniques of mathematical modelling have been used to predict times for temperature equilibration and microbial growth in food products during freezing and equilibration to frozen conditions. This approach, which is in its early days, represents an important advance in being able to predict the behaviour of microbes during frozen processing (Castell-Perez *et al.* 1989).

On the other hand, because of practical difficulties, there has been little investigation of the effects of the spoilage bacteria existing on foods, prior to freezing, on the quality changes occurring during frozen storage. Both the detection of injured microbes and the linking of quality change to microbial populations have tremendous practical importance whether the topic is the protection of public health or the retention of quality.

The long potential shelf-life of foods at deep freeze temperatures is limited by chemical not microbiological changes. At temperatures below -6°C to -10°C microbial growth does not occur and the progressive spoilage normally associated with it is absent. The important changes adversely affecting sensory quality of frozen foods will be chemical and physical ones, including oxidation and desiccation. Commercially, such deterioration may be controlled in a number of ways, for example by the use of blanching, the addition of chemical anti-oxidants or the choice of functional packaging techniques limiting the access of oxygen to the product and the loss of water from it. The microbiological and chemical quality of the food prior to freezing also critically influences the type and extent of the quality changes taking place during frozen storage. These changes are generally controlled by the careful choice of raw materials, storage conditions prior to processing and manufacturing techniques.

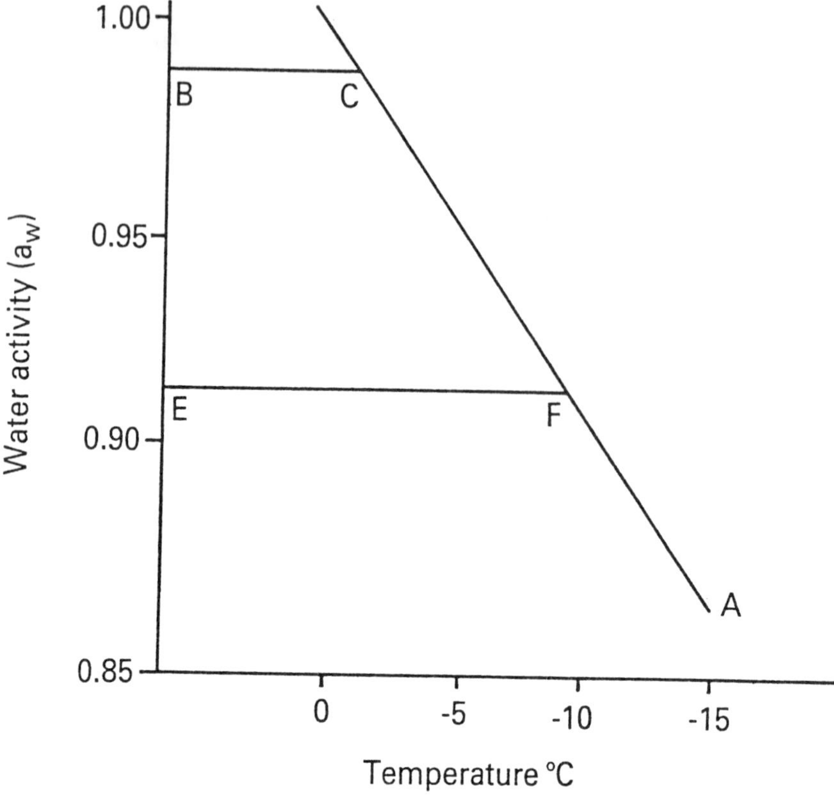

Fig. 2.1 The effect of water activity and temperature on freezing point, shown for water containing no solutes (line CA), a high water activity food (line BCA) and a lower water activity food (line EFA).

Microbial Growth at Low Temperatures

The link between temperature and microbial growth rate is a complex one which has been empirically determined as an inverse square root or a non-linear Arrhenius relationship (Broughall and Brown 1984). As the temperature falls, growth rate is reduced and the lag phase may be extended until the minimum limit is exceeded and growth ceases. Some microorganisms will cease growing at 0°C or higher temperatures, whilst others will continue growing well below 0°C, below the freezing point of the food. Growth rates at sub-zero temperatures are very slow and generation times may exceed about 100 h.

The large differences in minimum growth temperature (10°C to -10°C) between microorganisms commonly associated with food (mesophilic and psychrophilic bacteria, yeasts and moulds) suggest that several different mechanisms are involved in limiting growth above 0°C, or eventually halting it at sub-zero temperatures.

The Cessation of Growth

Growth may be stopped well above the freezing point of the food, if the direct effect of temperature on metabolic rate causes maintenance energy requirements to exceed energy production (for example if the cell is stressed by low pH or some other means) or if bottlenecks in the biosynthesis of macromolecules limit the cell's ability to divide. The differential inhibition of enzyme activities at low temperatures can change, and may reduce, the efficiency of metabolic pathways, leading to the increased accumulation of end-products at temperatures around the growth minimum. This would be of particular importance in foods, such as raw meat and fish, which may be subjected to prolonged chill storage prior to freezing, allowing substantial microbial populations (e.g. 5×10^7/g) and metabolite concentrations to develop before processing and freezing. In raw meat, stored chilled under aerobic conditions, spoilage populations are often dominated by microorganisms (such as Pseudomonads) producing extra-cellular lipases and proteases which remain active and can cause continuing flavour changes during storage at frozen temperatures (see below).

Most foodstuffs contain low molecular weight solutes, such as salt and sugars, which depress freezing point and lower water activity. The graph (Fig. 2.1) shows this relationship. Once the freezing point has been passed, progressively more water is converted to ice and the water activity in the remaining unfrozen water is reduced. On the graph, the line BCA shows the freezing characteristics of a high water activity food (a_w = 0.98) and the line EFA shows them for a lower aw food (a_w = 0.91).

Therefore as the freezing point is approached at sub-zero temperatures, microorganisms not only approach their minimum growth temperature, but they are subjected to a progressive increase in the concentration of solutes in the unfrozen water, which is probably in osmotic equilibrium with the cell contents, and is used for metabolite diffusion and exchange. Strong inhibitory interactions occur between low temperature and low water activity, and growth is slower or ceases, at a higher temperature than if either factor were acting alone (Ingram and Mackay 1976). It is well known that the types of microorganism able to grow at low water activities will also grow in foods at sub-zero temperatures. Bacteria, yeasts and moulds are able to grow at successively lower water activities and the common food spoilage bacteria cease growing at about -5°C or a_w 0.95. Where a food product contains a mixed flora and is stored at temperatures near to the limits of growth, small changes in temperature can make substantial differences to relative growth rates of the microorganisms present and hence to the composition of the flora developing during storage.

The Selective Effect of Low and Sub-Zero Temperatures

Changes in the composition of the naturally occurring spoilage flora of an experimental minced beef product are an example of the selection of different species by decreasing temperature (see Table 2.1). Over the temperature range +5°C to -7°C or -8°C, the time taken for the microbial population to increase from 10^4/g to around 10^7/g increased from between three and four weeks to about 12 weeks. As the storage temperature was reduced, so different components of the contaminating flora became most numerous. Above 0°C the normal Gram negative became most numerous, at 0°C their growth rate had been slowed sufficiently for the Gram positive species to form a significant proportion of the population (the difference in count being less than 1000-fold). At sub-zero temperatures the growth rate of the bacteria had slowed sufficiently for the yeasts and moulds to be present in detectable numbers and with lower temperatures and increasing storage period they grew to become the only detectable species.

Table 2.1. The growth of various types of microorganism in an experimental minced beef product during frozen storage (-18°C) in air

Storage temperature	Storage period (weeks)			
°C	2	4	8	12
+5	Gm-	spoiled	-	-
+2	Gm-	Gm-	spoiled	-
0	Gm- and Gm+	Gm- and Gm +	Gm- and Gm +	-
-2 (aw=0.98)	Gm+, Gm- and Y	Gm+, Gm- and Y	Y	Y and M
-5 (aw=0.95)	Gm+ and Gm-	Gm+, Gm-, Y and M	Y	Y and M
-7/-8 (aw=0.92)	Gm+, Gm- and Y	Gm+, Gm-, Y and M	Y and M	Y and M

Gm+ = Gram positive species
Gm- = Gram negative species
Y = Yeasts
M = Moulds

Notes

Yeast detectable in -3 dilution, if not overgrown
Freezing point of meat -5°C to -8°C

Determination of the water activity in the product at sub-zero temperatures was done by removal of the surface ice and measurement of the Erh/aw of the remaining product using a Novasina and the freezing point (-7°C to -8°C) was determined using differential scanning calorimetry.

Characterisation of the microbial types recovered showed that the typical water activities allowing growth (at 25°C to 37°C) were 0.97 for the Gram negative species, 0.94 for the Gram positive species, 0.85 for the yeasts and 0.82 for the moulds; although these minimum values would probably be increased at lower temperatures.

Freezing

The injury caused to cells by freezing is an important aspect of the microbiology of frozen foods. Microorganisms differ in their sensitivity to freezing and their responses range from no injury, through sub-lethal injury to death. After freezing, some microorganisms, for example *Listeria monocytogenes*, may have their ability to initiate growth under thaw conditions substantially reduced by injury (Golden *et al.* 1988).

The rate of freezing, the storage temperature and temperature fluctuations during storage play a major role in determining the extent of sub-lethal injury and death. The normal cultural techniques used in microbiology do not allow the clear separation of the effects of freezing and thawing, because the parameters used to define viability and injury are linked to growth i.e. lag period, growth rate and nutrient requirements etc. For this reason most investigative work has been done on cells which have been subjected to a freeze/thaw cycle. In frozen foods the interaction of all these factors are likely to be important in determining the ability of cells to resume growth when favourable conditions return.

Freezing Rate

Freezing processes may be roughly classified according to the rate of freezing. Typical rates are:

Slow freezing - around 1°C to 10°C/h
Commercial freezing between 10°C and 50°C/h
Rapid freezing - above 50°C/h

Food will always freeze from outside towards the centre. The freezing rate will be a function of the size and surface area of the unit being frozen, its thermal conductivity and the gradients of temperature existing within it and the temperature differences at its boundaries. Hence it is likely that any microbial contaminants distributed within the food will have been subject to a range of freezing rates, and those on exposed surfaces additionally to desiccation. Any practical assessment of the effects of freezing on populations of microorganisms must take into account the different types of damage resulting from this variation.

The Effects of Freezing

Freezing has two major effects on the intra-cellular water and the water surrounding microorganisms. Firstly, the water is converted to ice crystals, whose size will be determined by freezing rate. Secondly, as water is transformed to ice, solutes will be progressively concentrated in the remaining unfrozen water. Slow and rapid freezing have opposite effects on crystallisation and solute concentration.

At slow freezing rates large ice crystals are formed and the ice crystal front moves slowly through the food being frozen, concentrating solutes ahead of it. Microorganisms are exposed to high solute concentrations and large ice crystals will be formed mainly outside the cell.

At rapid freezing rates numerous small ice crystals will be formed both intra- and extra-cellularly and solutes will not be concentrated to the same extent as by slow freezing processes. Microbial cells will retain their normal size, although they may show distortion and loss of membrane integrity because the ice crystals formed are small enough to disrupt the architecture of the membrane.

Microbial cells from commercially frozen foods may show the damage typical of both rates of freezing, but often there is little evidence of dehydration or membrane damage and the cells are less damaged than those subjected to either slow or rapid freezing.

The damage caused to microorganisms by intra-cellular freezing will depend partly on the rate of passage of water and solutes through the cell membranes (influencing osmotic dehydration of the cell contents and reducing the intra-cellular water activity a_w) and partly on whether the ice crystals formed within the cell cause damage to the structure and organisation of the cell. For example, lesions in the cell membrane will reduce the cell's ability to retain low molecular solutes and maintain its internal environment. The effects of such damage will not become evident until the cell is thawed and conditions become suitable for the initiation of growth. There is poor agreement from published experimental data on whether thawing rates have any significant interactions with the damage already caused by freezing (Ingram and Mackey 1976; van Schothorst 1976).

During storage, temperature fluctuations will cause the movement of solutes, the growth of ice crystals and possibly the loss of water from the product by sublimation. All these factors will increase the damage to cells.

Factors Influencing the Sensitivity of Micro-organisms to Freezing

Besides freezing rate, other factors may affect the sensitivity of microorganisms to freeze damage. Species differences, different growth conditions or the stage of growth will cause different responses to freezing. Some cells will survive unharmed, some will be injured during freezing and frozen storage, whilst others will be injured only when frozen under certain conditions. Actively growing cells are more sensitive to freeze damage than stationary phase cells. Bacterial spores are the form most resistant to freeze damage.

The spores and vegetative cells of yeasts and fungi survive freezing and frozen storage well.

Generally the Gram positive cocci, such as micrococci and streptococci, are resistant to freezing and survive frozen storage better than the Gram positive rods. Practical evidence of this is that the species most commonly isolated from environmental debris in deep freezes are Gram positive cocci. In contrast to this the Lactobacilli commonly used as starters for yoghurt and cheese manufacture in the dairy industry can suffer a four to six log reduction in numbers during frozen storage and handling, prior to inoculation into milk.

Many Gram negative species are damaged or killed by freezing. Their level of survival and susceptibility to damage is variable and although they are considered sensitive to freezing, this cannot be used to guarantee their absence in frozen products. For example Enterobacteriaceae can be recovered from frozen foods after storage for several years. Many of the methods now used to determine the numbers of Enterobacteriaceae in frozen foods, recognise their sensitivity to damage and include a step allowing repair of the cells prior to exposure to inhibitory or selective agents. The inability of freeze-damaged cells to grow on media containing these agents can lead to under-estimation of their numbers or over-estimation of their rate of death or the lethality of a particular process. Such errors in methodology have done much to perpetuate the myth of cold-store kill.

Storage temperature exerts a major effect on sub-lethal injury. Higher sub-freezing point temperatures (-2°C to -5°C) are generally more injurious to cells than lower temperatures (below -10°C). Salt concentration is also a critical factor, high levels contributing to osmotic dehydration and increasing damage during slow freezing. Many common food components (such as sugars, peptides and glycerol) can protect microorganisms from freeze damage (cryoprotectants), and some workers have found that the effect is not to reduce the extent of injury but to reduce the percentage of cells killed by a certain freezing treatment (Olson and Nottingham 1980).

Low pH accelerates death and increases sub-lethal injury rates during frozen storage (Georgala and Hurst 1963). Because measurement of pH at frozen temperatures is not possible, experiments leading to this conclusion have related the results to food pH before freezing, although it is likely that freeze concentration and the precipitation of salts will alter the actual pH around the microorganisms.

Sub-Lethal Injury

Cells that are not killed by freezing may be sub-lethally injured and such injury is reversible, repaired cells regaining the characteristics of normal cells. Injured cells lose viability if exposed to environments causing stress or containing inhibitory agents, but given suitable temperatures and nutrients most freeze injured cells will regain their original characteristics within several hours. The consequences of freeze injury are not transmitted at cell division, indicating that freezing does not cause permanent changes to the cell's genetic material. The sites of cellular damage and the mechanisms of repair have been identified and reviewed in some detail (Ray 1986).

One of the commonly observed effects of sub-lethal injury is leakage from the cell of low molecular weight cellular material including peptides and amino acids, indicating damage to the integrity of the cell membrane. Where large populations of microorganisms are present, this leakage of low molecular weight substances is thought to provide some degree of cryo-protection.

Injured cells often have increased nutritional requirements, impaired membrane permeability and nutrient transport characteristics and reduced resistance to environmental stress (for example low pH). Sensitivity to membrane-active inhibitory agents (such as the surface-active compounds used as selective agents in microbiological media) is increased. This has been taken to indicate that freezing causes major conformational and functional changes in the cellular structures controlling membrane integrity and permeability. This results in the movement of ions and low molecular weight substances to and from the cell and, secondly, to the mechanisms for coding and transcribing genetic information and the synthesis of proteins.

Repair and Resuscitation

Repair of freeze damaged cells occurs during the lag period. There is intense metabolic activity involving RNA and ATP synthesis and reorganisation of membrane components, such as lipopolysaccharide, without accompanying growth. The *de novo* synthesis of proteins and DNA has not been detected during repair from freeze damage. Many techniques have been developed to maximise the numbers of damaged cells recovered after freezing (Ray 1986; van Schothorst 1976 and Oscroft *et al.* 1987), but there is a lack of general agreement on optimum conditions for recovery of particular microorganisms or food products, reflecting the wide range of damage caused by freezing and the variety of conditions needed for successful repair. For example both nutritionally rich and minimal media have been successfully used to encourage repair (van Schothorst 1976).

Many studies on freeze-injury and its repair have been done using artificially contaminated foods, but the natural contaminants of commerically frozen food have received little study. In part this may be due to the difficulty of obtaining samples whose freezing and storage history can be adequately defined. Also the use of naturally contaminated foods raises another problem - one of obtaining accurate results to form the basis of quantitative conclusions and recommendations.

The microbial population in many types of foods is log normally distributed (Kilsby and Pugh 1981). In naturally contaminated foods, (unlike broth cultures) the spread of counts (i.e. the standard deviation) is often very large. Because the microorganisms are not uniformly distributed, existing as microcolonies - unless the food has been mixed or homogenised. The accuracy of the estimate of the population size obtained by sampling the food at a low level (five to ten samples/batch) is so low that any small change in count resulting from a particular resuscitation treatment can be hidden by the variability of the original distribution.

A study on the recovery of damaged or injured microorganisms from commercially frozen food has been done by Oscroft *et al.* (1987). They investigated the efficiency of Plate Count Agar (PCA) at recovering injured bacteria from a range of frozen foods and compared the effects of solid-phase resuscitation, salt level and incubation temperature on the numbers recovered. Examination of their data relating to one particular commodity, minced beef, highlights the problem of accounting for the variation attributable to the log normal distribution of microorganisms, when looking for small effects attributable to different recovery methods (Table 2.2).

Table 2.2. The effect of various resuscitation techniques on the recovery of the natural microflora of frozen minced beef (from Oscroft *et al.* 1987)

Medium used	Incubation temperature	Maximum and minimum % change in count
Peptone buffer diluent	20°C	+18% to +178%
Minerals modified glutamate	20°C	+15% to +57%
Trypticase soy agar	20°C	- 35% to +50%
PCA (non resuscitation)	20°C	- 58% to +35%

Unfrozen minced beef - log count/g = 7.73

The percentage increase or decrease, after freezing and resuscitation, is based on the count in the unfrozen minced beef.

Their experiments examined the effects of a number of different factors, including four different types of resuscitation media, on the numbers of microorganisms recovered (Table 2.2). Using incubation at 20°C, they found that different resuscitation techniques could increase (+178%) or decrease (-50%) the numbers of microorganisms recovered. If the distribution of their data on the numbers of microorganisms recovered from mince by the various resuscitation treatments is assumed to be log normal (Mean log count = 7.79; Standard deviation = 0.34), it can be seen that the variablility attributable to all their different treatments is lower than that occurring within the batches of naturally contaminated mince (Mean log count = 6.9; Standard deviation = 0.54) examined by Kilsby and Pugh (1981) using a single recovery technique. Against such a background of variability it is very difficult to draw confident conclusions on the benefits of any particular method, unless particular attention has been paid to the design of experiments to take account of the variability in the materials examined.

Microbiology and Sensory Quality

The harvesting, handling and storage procedures used for foods destined for freezing can determine the contribution of microbial enzymes and metabolites to the type and extent of the quality changes occurring during frozen storage.

When storage periods are short, processing is rapid and may involve a pasteurisation step, the small amount of microbial growth possible before freezing will therefore have a negligible effect on the quality changes. An example of such a rapid process is the commercial harvesting and processing of vegetables for freezing, which because of its speed (a few hours from harvesting to freezing) provides limited opportunities for microbial enzymes and by-products to build up. Quality loss in frozen vegetables is likely to be caused by residual enzyme activity in the vegetables themselves, especially from any enzymes surviving blanching. It may also be caused by other factors, such as desiccation or mechanical damage during processing.

Poor process hygiene, allowing recontamination after the blanching stage may result in high counts on blanched vegetables. Generally contaminants picked up in this way will not have been active on the vegetables long enough to cause quality changes. High counts should therefore not be automatically associated with quality problems, unless the origin of the bacteria is known and the residence times and temperatures involved are long enough to allow significant growth. Little is known about the effects, during storage, of any extra-cellular microbial enzymes picked-up during post-blanch handling.

Where the time between harvesting and freezing is prolonged and may involve chilled storage, microorganisms can grow prior to freezing (Cheuk 1988; Tsao *et al.* 1988; Ward and Baj 1988). The enzymes and metabolites they produce have been shown to contribute to quality changes during frozen storage (Hall and Alcock 1987). For example during the chilled storage of raw meat, the spoilage population can become sufficiently large for it to make a recognisable contribution to the enzymic and metabolite composition of the meat. These by-products can cause progressive flavour changes during frozen storage. In practical terms, microbial populations growing to levels in excess of one to ten million per gram prior to freezing, can cause significant changes in products during frozen storage, although these numbers would not normally be recognised as being associated with spoiled raw materials (ICMSF 1986).

The growth of spoilage bacteria, to levels around ten million/gram in raw beef mince stored at room temperature prior to freezing, has been found to cause the progressive development of off-flavours, greasiness and offal notes, assessed by the sensory panellists as a Just-noticeable-difference (JND), during eight months frozen storage at -15°C. Where the average count had reached 100 million/gram the initial quality was so low that further quality loss during frozen storage could not be detected (Hall and Alcock 1987). These authors also concluded on the basis of their experiments, involving pre-storage of the materials at room temperature, that microbial enzymes could play a significant role in quality changes in meat and vegetables during frozen storage. They found that in commercially produced peas, microbial populations below one million per gram did not contribute to quality change, but in the inoculated, pre-stored samples, such a level did cause quality deterioration during frozen storage. This highlights the importance of determining the origin of contaminants and their opportunities for producing enzymes and by-products when assessing the importance of contamination.

Two reasons suggest that it is not possible to use single, or even very low numbers of microbiological samples to reliably predict quality changes during frozen storage or to precisely specify the levels of contamination likely to cause spoilage. Firstly, the log normal distribution of bacteria, which often has a high variability in unprocessed materials, means that their potential for quality change during subsequent storage will not be reliably indicated by an average or single count. A single value provides no indication of the range of counts and any very highly contaminated areas present are unlikely to be detected by low sampling rates, although they will have an over-riding influence on quality. Secondly, because the perception of spoilage will differ from person to person and also depend on the character of the product, it is difficult to provide general microbiological limits covering all raw materials.

Reliable data, suitable for forming the basis of specifications, can only be obtained from experiments designed to link microbiological quality (and the distribution of qualities in the raw material) to perception of the product by the customer.

Unfortunately, because such experiments are difficult to organise and the results are only applicable to particular product groups, they are infrequently undertaken and published - although they would make an important contribution to the development of standards in legislation (see Garrett 1981).

Other Compounds Formed During Pre-Storage

In scombroid fish, such as herring and tuna, the formation of biogenic amines, such as histamine from the histidine already present in the tissues, is thought to accompany the growth of some types of spoilage bacteria during storage (Taylor *et al.* 1979; Taylor 1988; Park *et al.* 1988). Sometimes this storage may be prior to freezing. The formation of this toxin can be controlled by rapid chilling which prevents the growth of decarboxylating bacteria such as *Proteus*, *Hafnia* and *Klebsiella*. Biogenic amines will not be destroyed by freezing and may persist in frozen foods, but there is no evidence that their formation continues during frozen storage. With the increasing use of fish from tropical climates, where the achievement of chilled temperatures during handling and transport is difficult, this may become an increasingly important topic.

Conclusions

There are numerous good reasons for wanting to estimate accurately the size and composition of the microbial populations in frozen foods. Freezing certainly stops any microbial growth during storage, so that the risks of hygienically manufactured frozen foods containing harmful numbers of microorganisms is low. Some frozen foods, however, are designed to be eaten without further cooking and others may not necessarily be heated to pasteurising temperatures by all customers; the absence of microorganisms of concern is therefore of importance. Hence there is a need for accurate and reliable microbiological methods for use as part of quality assurance systems.

Freezing and frozen storage injures many types of food-borne microorganisms, as described above. Injury leading to a restriction of the ability of microorganisms to grow is one of the major causes of detection inaccuracies. The rate of freezing and the composition of the medium surrounding the microorganisms will affect the type and extent of injury. Consequently, a single microbiological method cannot be expected to give consistent and predictable recoveries for all frozen foods. Knowledge of the cellular consequences of freezing has led slowly to a logical review of recovery methods. Much published work recommending various techniques for recovery only summarises experience or is based on comparative experiments. Rarely does it link knowledge of the physiological basis of freeze-injury with the mechanisms of repair thought to be stimulated by various media. Advances in recovery will only come from a clear understanding of the physiological basis of injury and of repair. The use of this knowledge to design recovery media, which alleviates or stimulates particular aspects of the metabolism of injured cells, therefore allows normal growth to take place.

Most importantly, any changes in the efficiency of recovery should be reflected in the review of microbiological limits and specifications. Any improvement in techniques leading to a substantial increase in the numbers or range of injured microorganisms recovered, represents an effective tightening of limits.

Apart from microbiological safety issues, the quality of the food before freezing often determines the quality changes occurring during storage. There is a limited understanding of what, if any, effects microorganisms exert on quality or whether the impact of microbial populations is different in frozen foods (shelf-life six months to one year) and chilled foods (shelf-life about one to two weeks). Such experimental work entails multi-disciplinary, large scale experiments with real products and packaging systems and this is rarely undertaken. Unfortunately, this lack of understanding does not stop food manufacturers and retailers imposing arbitrary limits, without a knowledge of the extent of customer protection or the commercial penalties incurred. This is an area where effective experimentation could give the customer and industry real benefits.

References

Broughall JM, Brown C, (1984) Hazard analysis applied to microbial growth in foods: development and application of three-dimensional models to predict bacterial growth. Food Micro 1, 13-32

Castell-Perez ME, Heldman DR, Steffe JF (1989) Computer simulation of microbial growth during freezing and frozen storage. J Fd Proc Eng 10:249-268

Cheuk WL (1988) Effect of on-board handling and processing techniques on the quality of sharks from the Gulf of Mexico. Diss Abs Int B47 (1) 15

Garrett ES (1981) Microbiological standards, guidelines and specifications and inspection of sea food products. Fd Technol 42 (3) 90-93, 103

Georgala DL, Hurst A (1963) The survival of food poisoning bacteria in frozen foods. J Appl Bacteriol 26 346-358

Golden DA, Beuchat LR, Brackett RE (1988) Inactivation and injury of Listeria monocytogenes as affected by heating and freezing. Fd Microbiol 5 17-23

Hall LP, Alcock SJ (1987) The effect of microbial enzymes on the quality of frozen foods. Fd Microbiol 4:209-219

ICMSF (1986) The application of variables plans. Chapter 8 in Microorganisms in foods, 2: Sampling for microbiological analysis: Principles and specific applications. Blackwell Scientific Publications, Oxford

ICMSF (1988) Food processing 11.9 Freezing pre-cooked shrimps and prawns in Microorganisms in foods, 4: Application of the hazard analysis critical control point (HACCP) system to ensure microbiological safety and quality. Blackwell Scientific Publications, Oxford

Ingram M, Mackay BM (1976) Inactivation by cold. 111-151 In: Skinner FA, Hugo WB (eds) Inhibition and Inactivation of Vegetative Microbes published for the Society of Applied Bacteriology by Academic Press

Kilsby DC, Pugh ME (1981) The relevance of the distribution of microorganisms within batches of food to the control of microbiological hazards from foods. J Appl Bacteriol 51:345-354

Olson JC Jr, Nottingham PM (1980) Temperature. 1-37 In: Microbial Ecology of Foods, Vol 1 Factors affecting life and death of microorganisms. ICMSF. Published by Academic Press

Oscroft CA, Alcock SJ, Clayden JA (1987) Recovery of sub-lethally injured bacteria from frozen foods. Fd Microbiol 4:257-268

Park HY, Oh HS, Lee EH (1989) Frozen storage stability of seasoned anchovy products. Korean J Fd Sci Technol 21:536-541

Ray B (1986) Impact of bacterial injury and repair in Food Microbiology: its past, present and future. J Fd Prot 49:651-655

Taylor SL, Guthertz LS, Leatherwood M, Leiber ER (1979) Histamine production by Klebsiella pneumoniae and an incident of Scombroid fish poisoning. Appl Env Microbiol 37:274-278

Taylor SL (1988) Marine toxins of microbial origin. Fd Technol 42 (3) 94-98

Tsao CY, Hwang BS, Jiang ST (1988) Effects of pre-chilling on the quality of frozen shrimp. J Chinese Agric Chem Soc 26:156-164

van Schothorst M (1976) Resuscitation of injured bacteria in foods. 317-325 In: Skinner FA, Hugo WB (eds) Inhibition and Inactivation of Vegetative Microbes published for The Society of Applied Bacteriology by Academic Press

Ward DR, Baj NJ (1988) Factors affecting the microbiological quality of sea foods. Fd Technol 42 (3) 85-89

Chapter 3

Freezing in Polymer-Water Systems

G. Blond and B. Colas

Introduction

The properties of frozen aqueous systems are determined mainly by the proportion of unfrozen water, the ice crystal size and the physical state of the non aqueous components.

The shelf-life of products stored at low temperature is dependent on the temperature because the unfrozen fraction may be a liquid or a glass phase according to the temperature. The preservation of the frozen product characteristics is bound to the chemical and biochemical reaction possibilities which are strongly decreased if the freeze-concentrated phase is a glass (Levine and Slade 1989, Simatos and Blond 1990).

A qualitative interpretation of the freezing behaviour of such complex systems as biological or food systems can be obtained from the study of model systems. It seems that the presence of macromolecules modifies the behaviour of the water during the cooling process. Numerous stabilisers have been used, for example in ice creams, with the aim of inhibiting ice crystal growth, but this claim is poorly supported by sound data.

Besides these potential applications, the knowledge of the influence of macromolecules on the freezing of the water may also provide basic information on the properties of the water in these solutions.

This paper discusses the peculiar features of the crystallisation of water in macromolecular model systems with and without small solutes. It presents successively the physical state of water in frozen systems, the freezing point depression and the nucleation and ice crystal growth phenomena in polymer solutions. Biological frozen products being always complex systems, only a qualitative interpretation can be obtained from these studies.

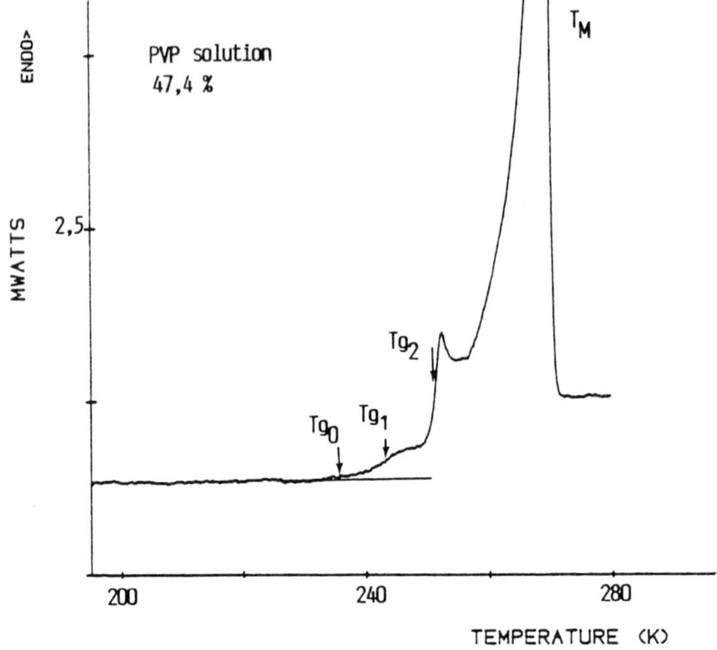

Fig. 3.1. DSC heating scan for a 47.4% PVP solution showing the thermal transitions reported in the state diagram. The stabilisation of the transitions was obtained by a first warming to 250 K in the calorimeter (cooling and heating rate:10 K/min).

Physical State of Water in Frozen Systems

In a frozen system most of the water is transformed into ice but a significant water fraction can remain in an amorphous state. The questions are: which water fraction is "unfreezable"; at which temperature is the freeze-concentrated fraction transformed into a glass?

The physical state of water in frozen solutions may be illustrated by a state diagram which represents the various states in which a system can exist as a function of temperature and concentration. The data on the freezing behaviour, i.e. the first and second order phase transitions observed during the cooling and/or the subsequent rewarming of aqueous solutions as obtained by DSC experiments (Fig. 3.1), are combined to furnish a state diagram.

Taking the polyvinylpyrrolidone (PVP)-water system for illustration, we can distinguish the equilibrium from the non-equilibrium changes. This diagram was previously presented by Mackenzie and Rasmussen (1972) and by Franks *et al.* (1977). Fig. 3.2 represents the state diagram we constructed with our DSC measurements; the bold lines represent equilibrium curves and the dotted lines indicate processes under kinetic control.

The curve M represents the temperature at which ice begins to separate when a solution of concentration lower than 60% is cooled under equilibrium conditions. When ice appears the concentration of liquid phase increases. PVP, as most biological

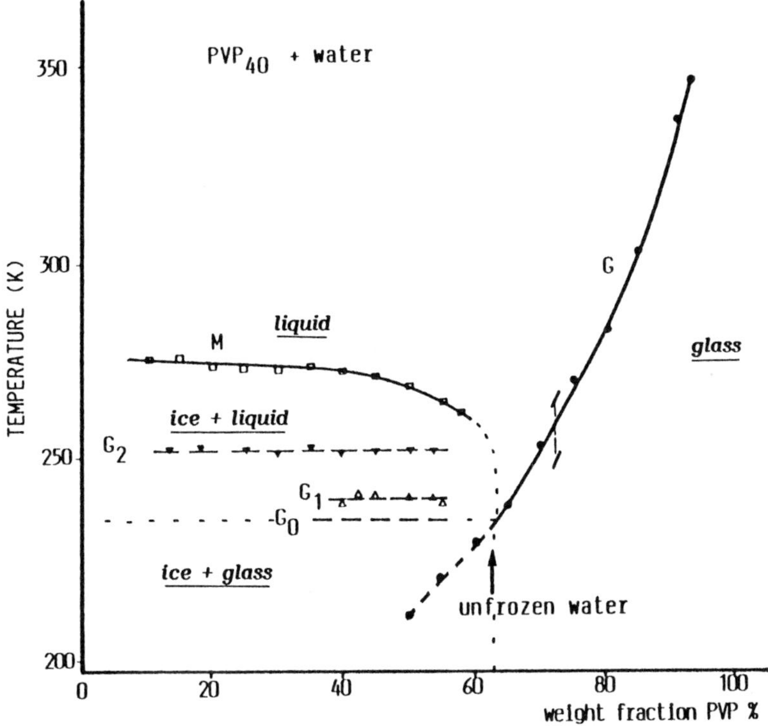

Fig. 3.2. State diagram of the system PVP-water. The underlined terms designate the phases which are present in samples in which all "freezable water" was actually frozen.

solutes, does not crystallise during cooling, the concentrated liquid phase is solidified as a glassy material when it reaches the G curve. The water remaining in the amorphous phase is often discussed as "unfreezable water".

The curve G represents the glass transition temperature as a function of PVP concentration.

The presence of the glass phase in the system is revealed on the thermogram by a change in the specific heat, the glass transition. However with the PVP solutions the DSC curves exhibit the complex features (Fig. 3.1) which are also observed with sugar solutions (Blond 1989).

On the thermogram three temperatures are indicated: Tg_0 at the onset of the glass transition; Tg_1 at the middle point of the first event and Tg_2 at the middle point of the second event, but which is not well defined because of its overlapping with the melting endotherm. Presently, it is certain that the glass transition begins at Tg_0 but uncertainty remains concerning the exact starting temperature of ice melting. This "double peak" feature may be considered as representative of a glass transition associated with enthalpy relaxation. Moreover the reason for the presence of this important enthalpy relaxation in samples containing ice phase is not known (Simatos and Blond 1990).

The comparison of thermograms obtained with different PVP solutions at concentrations lower than 60%, and, after either a slow cooling or a rapid cooling, followed by an annealing treatment at -23°C (250K), shows that "the glass transition event" always occurs at the same temperature. Thus it can be concluded that a glass of constant composition may be formed corresponding to a defined amount of unfrozen water.

The amount of unfrozen water can be calculated from the experimental melting enthalpy (Blond 1989). The linear regression of the melting enthalpy as a function of water content gives 0.58 g of unfrozen water per g of PVP; this content appears to be close to the intersection of the curve G with the line G_0 (Fig. 3.2).

The phase states present in systems in which the maximal ice crystallisation was obtained are indicated on the state diagram (Fig. 3.2). If the storage temperature is lower than Tg_0, the freeze-concentrated amorphous phase will be a glass; if it is higher than Tg_0, it will be a supercooled liquid phase. In certain cooling circumstances, e.g. rapid cooling, the amorphous phase may be more diluted than the maximally freeze-concentrated one. This glass transition temperature will be lower than Tg_0 (Tg if there is no freeze-concentration at all). In commercial storage the amorphous phase is always liquid since for most food products the temperature range corresponding to the glass transition of the maximally freeze-concentrated phase (and to the beginning of the melting of ice) appears to be from -20°C to -45°C (Rey 1964).

To obtain a glassy material with a system containing a freezable water fraction, the cooling rate must be very rapid and/or the concentration high because the viscosity of the liquid phase limits the nucleation possibility. During storage, the glass state is preserved with a temperature lower than the G curve.

Freezing Point Depression (FPD or ΔTm)

The freezing point depression is the temperature at which the first ice crystals can appear when cooling is carried out under equilibrium conditions.

The study of the thermodynamic properties of a complex system can give some information on the average properties of the macrosystem.

A brief review of thermodynamic principles indicates that for an ideal solution the depression of the chemical potential is proportional to the molar fraction of the solute and that the variation of the vapour pressure may be conveniently expressed by the Raoult's law:

$$P_w = P_w^o . X_w$$

with P_w^o vapour pressure of pure water,

P_w, vapour pressure of water in the solution,

and X_w, water molar fraction.

As a consequence, when a non-volatile solute is dissolved in the pure solvent, the colligative properties of the solution vary with the solute concentration. Among the colligative properties, the freezing point is the temperature at which solid and liquid phases coexist in equilibrium. When ice and an aqueous solution are in equilibrium, the chemical potential of water must be the same in both phases. As the chemical potential of water is depressed by the solute, the freezing point depression (FPD) of water in the solution can be calculated from the Raoult's law.

The FPD in a macromolecular solution is expected to be very small because of the large molecule weight of the solute. Unlike the small solutes, macromolecular components may have a comparatively weak influence but it is worth considering. Today very accurate cryoscopic equipment, initially designed for use in the dairy industry and for medical investigations on biological fluids, are available.

The solutions of macromolecules behave in a non-ideal way; the decrease of water vapour pressure, and consequently the FPD, are generally larger than would be expected from Raoult's law. The vapour pressure of water in solution results from three effects:

i the difference in the dimensions between solute and solvent molecules; the Flory-Huggins's equation expresses the corresponding deviation;

ii the solvation effect: specific interactions between the solute and the solvent can reduce the vapour pressure of the solvent, an hydration number characterises this effect; and

iii the solute-solute interactions.

In an actual system interplay among these various effects may be present and it is difficult to distinguish each of them in the resultant feature.

The analysis of deviations from Raoult's law, which can be observed in polymer solutions, provides information relevant to the understanding of the nature of such solutions because they are of necessity an indication of solute-water and solute-solute interactions.

The FPD increases strongly with the increasing concentration of the polymers. For example with Carboxymethylcellulose (CMC, Mw = 80000) it varies between 0.1°C and 0.4°C when the concentration increases from 2% to 9%. Similar results have been observed with polyvinyl alcohol (PVA), low methoxyl pectin (LM pectin) (Blond 1985). Mackenzie (1975) determined freezing points for various solutions (PVP and collagen solutions or more complex systems such as erythrocytes) at higher concentrations (30% to 60%). The highly non-ideal nature of the experimental curves was evident in each instance.

These observations are in agreement with the osmotic properties of PVP and Hydroxyethylstarch (HES) solutions in the concentration range 10%-50% as reported by Franks (1982). Solms and Rijke (1971) reported that the FPD of several polymer solutions, either aqueous or non aqueous, were much larger than the values predicted by the Flory-Huggins's equation when using the interaction parameter, the value obtained from vapour pressure measurements.

The non-ideality of polymer solutions increases with their molecular weight. This can be expressed by the empirical Norrish equation:

$$a_w = X_w \exp(-K.X_s)$$

with a_w, water activity,

X_w, water molar fraction,

and X_s, solute molar fraction.

For example with the polyethylene glycols (PEG 200 to 35000) the relation between K and the molecular weight M is:

$$K = 8.9 \,.\, 10^{-4} \,.\, M^{1.91} \text{ (Voilley \textit{et al.} 1989).}$$

This non-ideality is amplified when the viscosity of the system is increased (Blond 1985). Gelling induces a significant increase of the FPD: the truly elastic gel which is

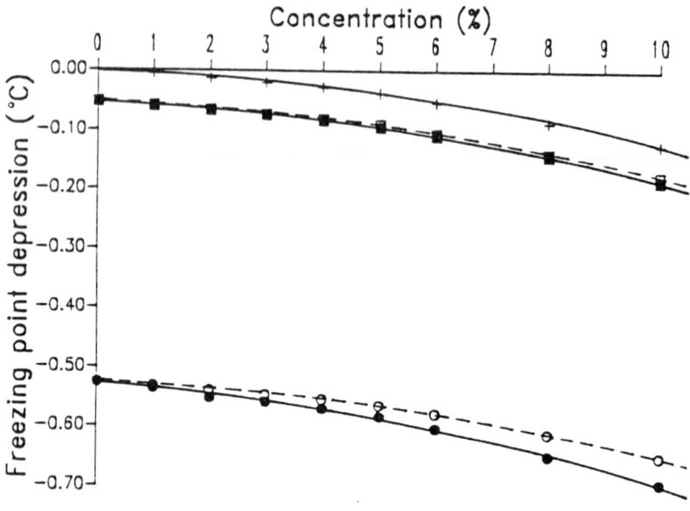

Fig. 3.3. Effect of a small solute addition (1%) on the freezing point depression of PVP solutions.
+PVP,□PVP+sucrose, O PVP+NaCl
open sign : calculated data
closed sign : experimental data

obtained by gelling polyvinyl alcohol with Congo Red shows a larger FPD than the one predicted by summation of the respective FPD of the two components. The structure of the gelatin gel obtained by ageing at 20°C results in a significant increase of the FPD (0.308°C after 24 h maturation instead of 0.137°C after one hour maturation for a 10% gel). In contrast, the viscoelastic gels obtained by ionic bridges (PVA + Borax or LM Pectin + $CaCl_2$) do not present the same effect, but it is difficult to know the free ion fraction in the gel and to take it into account in the calculations. Kuhn (1956) and Solms *et al.* (1971) have attributed the anomalous FPD observed in polymer gels to an impedence of the growth of solvent crystals. Another explanation could be that these systems behave like porous media where the vapour pressure of water is lowered by capillary phenomena, although stable capillary structures have not been identified in aqueous gels.

Often food systems are more complex and a better model should comprise different molecular weights. It is therefore interesting to study the effect of the addition of macromolecules to small solute solutions to know if competitive interactions could affect significantly their thermodynamic properties, particularly their FPD. Figure 3.3 (our results) shows the effect of an increasing amount of PVP in 1%-sucrose or -sodium chloride solutions. The shift between the calculated and the experimental curves increases with the polymer concentration, mainly with sodium chloride. Mackenzie (1975) published freezing point curves describing the behaviour of PVP + sodium chloride solutions (PVP/NaCl = 7.6/1.0 w/w). The difference between the calculated and measured curves suggested not only the non-additivity of the solutes contributions but an increased potential for water to bind with the polymer in the presence of salt. The same effect was shown by Korber and Scheiwe (1980) (with HES-sodium chloride-water systems) where an important variation of the liquidus curve with polymer concentrations was described.

Rey and Labuza (1981) obtained a similar effect with carrageenan + sucrose or potassium chloride, a greater interaction effect being measured with sucrose (Table 3.1).

Table 3.1. Influence of small solutes on the freezing point depression of a polymer solution

		Carrageenan 2%*		Avicel 1%+	
Concentration of added solute	ΔTsol	ΔTcal	ΔTexp	ΔTcal	ΔTexp
0			0.063		0.007
8.5% sucrose	0.480	0.543	0.755		
0.8% K Cl	0.424	0.487	0.520		
5% sucrose	0.280			0.288	0.283
10% sucrose	0.563			0.571	0.565
5% NaCl	2.889			2.896	2.862

Δ Texp : experimental data
Δ Tcal : calculated from the summation of each component value recalculated from * Rey and Labuza (1981) and + Chen *et al.* (1984).

With Avicel, a crystallised cellulose, Chen *et al.* (1983) found a negligible influence of the solutes (Table 3.1). This example illustrates the effect of water-solute interactions: the water-solute interactions would be limited by a crystallised material. The results suggest a small binding of sodium chloride on Avicel surface.

As early as 1938, Pyenson and Dalhes reported numerous measurements of FPD to determine the "bound water" in dairy products and polymers. Although their interpretations could be discussed, their results showed an increase of the non-ideality of the mixtures when macromolecules were added in the concentration range of 0.5% to 9% in a molar sucrose solution.

It is also important to note the influence of a polymer in a more concentrated medium or in a complex system. Some examples can be found in the literature. Muhr (1983) observed that the addition of a low concentration of food gums increased the FPD of concentrated sucrose solutions much more than it would depress the freezing point of water. Hoo and McLellan (1987) showed that pectin contributed significantly to the FPD of apple juice concentrates. In ice creams numerous stabilisers have been used to restrict the ice crystal growth (Willenger and Smith 1975), but no effect on the FPD was shown, perhaps because of their very low concentration (0.1% to 0.2%) as compared with that of the sweeteners in the mix.

Generally the non-ideality which is observed with macromolecular solutions increases strongly with their concentration. This effect is enlarged by the presence of small solutes. The examination of the FPD in polymer-solute systems suggests a steepening effect depending on the solute-polymer interactions.

When a product is frozen the interstitial liquid concentration increases rapidly in a non-ideal way. The water vapour pressure decreases significantly reducing the freezable water fraction at every temperature.

Nucleation

The FPD provides limited information because the thermodynamic equations alone cannot predict how the addition of many different solutes will change the water freezing process. Non-thermodynamic measurements are then required for a more informative description of the freezing process.

The development of the ice matrix is the result of two simultaneous events:

i the nucleation which represents the appearance of the ice nuclei;
ii the crystal growth, the process whereby the ice nuclei increase.

These two events are strongly interrelated because of the important effect of temperature on both processes.

Concerning the nucleation, there are two possibilities:

i heterogeneous nucleation: where a suitable interface might be provided by a foreign particle; or
ii homogeneous nucleation: where clusters of water molecules form spontaneously with sufficient size; these nuclei are the necessary interface for crystallisation.

Although, heterogeneous nucleation is the more likely process, homogeneous nucleation was widely studied because it is more easily accessible to theoretical analysis. For a large number of samples (with small droplets for example) the theory allows the calculation of the nucleation rate J which is the probability of having a nucleus per unit time and per unit volume. The crystallisation probability for a sample volume V is given by J.V for a given temperature; this can be illustrated by the data collected by Bigg (1953). The theory also allows the calculation of the cumulative nuclei concentration which is the number of active nuclei present in the unit volume at all temperatures higher than the specified temperature. This allows the determination of the temperature range where almost all droplets crystallise.

The relations are similar for both homogeneous and heterogeneous nucleation. The potential barrier is lower with heterogeneous nucleation and the transformation occurs at a higher temperature.

The main feature of crystallisation is its stochastic character, i.e. samples which are apparently identical will not crystallise at the same temperature during the cooling process. It is only possible to determine a probability of transformation by unit time and a statistical method is necessary to analyse the experiments. One must not also forget that the so-called nucleation temperature is in fact the temperature at which nucleation becomes sufficiently rapid for its experimental measurement.

It is commonly believed that hydrophilic gums influence ice formation in frozen confectionery products owing to qualitative observation of their effects on the ice crystal texture. Few experiments have been performed on this form of homogeneous nucleation but in most investigations heterogeneous nucleation cannot be avoided. The experimental results are described here in qualitative terms only.

Homogeneous Nucleation

To ensure that only homogeneous nucleation is measured, the analysed liquid is emulsified, making possible the observation of a great number of microsamples. Just as solute concentration affects colligative properties of solutions, so it can modify the nucleation temperature of ice in aqueous solutions (homogeneous nucleation temperature of water $T_h = -40°C$).

Investigations on some simple compounds have shown that in practice:

$$\Delta T_h = 2\ \Delta T_m \text{ (Mackenzie 1977)}$$

but deviation occurs for macromolecules (PEG and PVP) where ΔT_h is roughly five times greater than ΔT_m. The ability of polymer material to prevent ice nucleation in a non colligative manner is not a well resolved phenomenon.

These first results were complemented by Michelmore and Franks (1982) and Franks *et al.* (1983,1984) who showed that the presence of a solute especially with a high molecular weight such as PEG or HES, markedly affected ice nucleation. These authors explained the decrease of the homogeneous nucleation rate by an increase in the free interfacial energy between the ice and the aqueous phase. The effects could be more pronounced for HES solutions than for PEG ones. Franks *et al.* (1987) indicated that PVP and antifreeze proteins had the same influence on T_h.

Muhr *et al.* (1986) studied the influence of polysaccharide addition in sucrose solutions; their results showed that the effect of small quantities (0.3%) was weak and that this effect was masked by the experimental scatter.

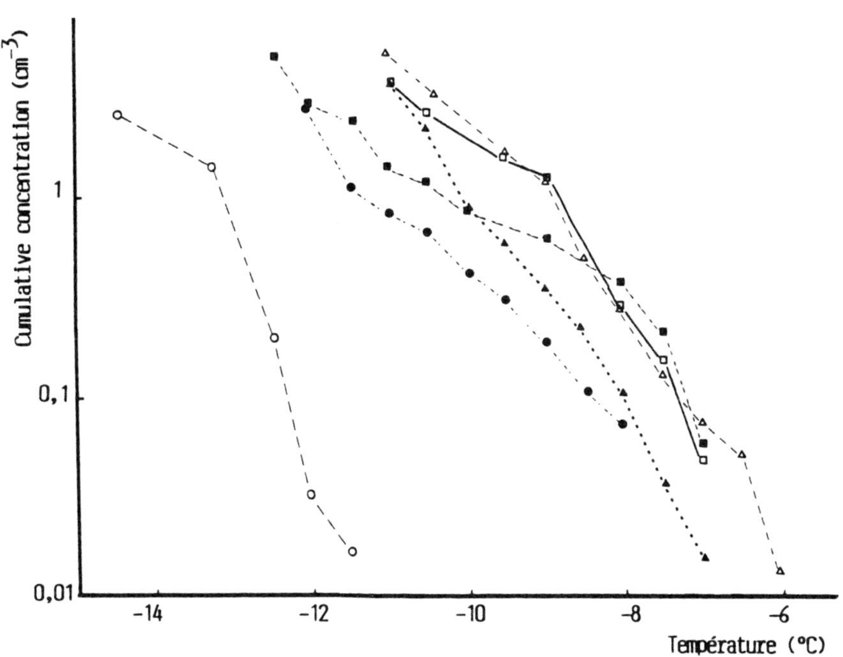

Fig. 3.4. Cumulative nuclei concentration as a function of temperature for O water, ● gelatin gel, ▲pectin gel, Δpectin solution, ■CMC high viscosity,□CMC low viscosity (sample size:2 cm^3) (Blond 1986).

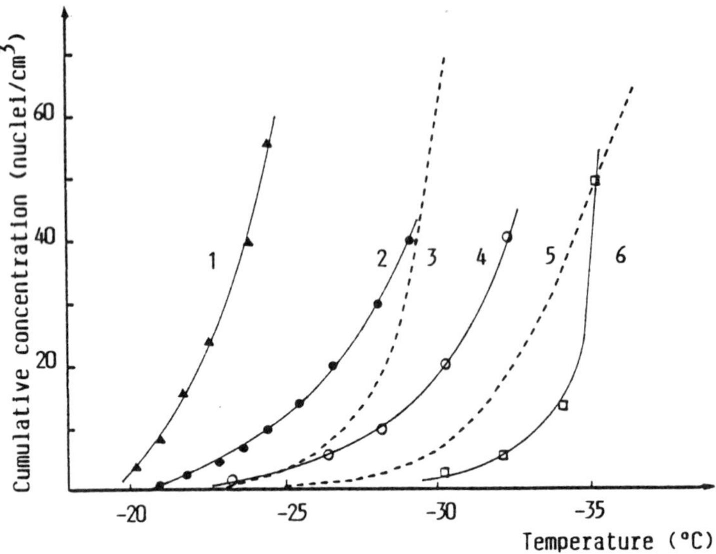

Fig. 3.5. Effect on heterogeneous nucleation of dissolving the specified stabilisers (1g in the stock 0.75 g/ml sucrose solution 150 cm^3).
1: manucol DM, 2: manucol DF, 4: sodium pectate, 6: xanthan gum, 3: tap water, 5: distilled water. (redrawn from Muhr *et al.* 1986).

Heterogeneous Nucleation

Parungo and Wood (1968) have studied the effects of macromolecules in order to know in which way they promote spontaneous freezing of ice from undercooled drops in the atmosphere. In agar solutions at very low concentration (0.01% to 0.1%) they reported a nucleation temperature higher than the T_h of water. The enhancing influence of foreign nuclei in macromolecule solutes was shown in all experiments. Using a droplet freezing statistical method, Reid (1983) demonstrated that food polymers (CMC, guar, carrageenan) increased the heterogeneous nucleation rate of water. The same effect was obtained at low concentrations (2%) with macromolecular dispersions divided into 2 cm^3 samples (Fig. 3.4).

The addition of food gums in sucrose solutions resulted in a large range of nucleating abilities (Fig. 3.5): some polymers such as methylcellulose or xanthan gum were found to have a slightly inhibitory effect; on the contrary a substantial increase in the nuclei number was observed for sodium alginates.

It is interesting to note that the gelification of polymeric systems, pectin for example (Fig. 3.4) decreases the nucleation temperature. But it is well known that the water molecules in aqueous gels suffer very little inhibition of their motional properties; the polymer network might disperse the water in small samples and then decrease the probability of heterogeneous nucleation at a given temperature. Experimentation demonstrates that common food grade polymers increase the heterogeneous nuclei population in water; many nuclei could be associated with the polymer because of its poor dispersion. Moreover, nucleation could also result from the presence of foreign particles.

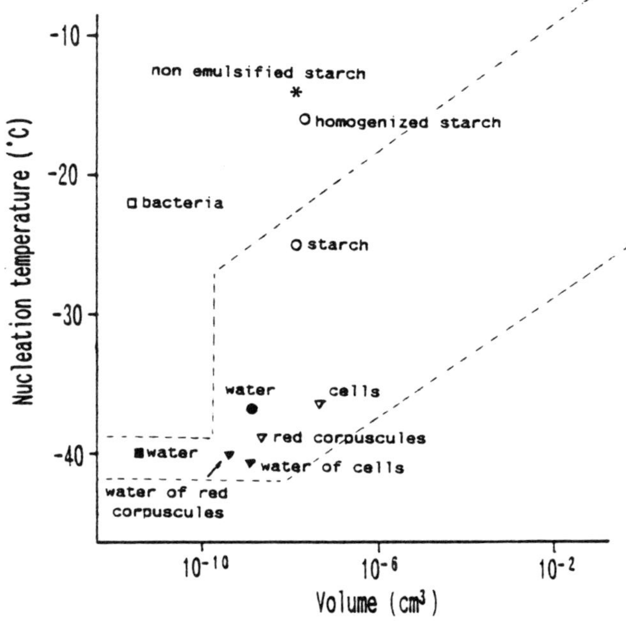

Fig. 3.6. Nucleation temperature of different emulsified systems and corresponding liquid phases as a function of sample size.
▽ data from Franks *et al.* (1983); □ data from Clausse and Jolivet (1989);
our results: O native starch and homogenised starch emusified; * starch suspension at critical concentration; dotted lines: limits of data on undercooling of water samples collected by Bigg (1953).

The use of an emulsifying technique to preserve cells from freezing damages was proposed by Mathias *et al.* (1984), but the study showed that homogeneous nucleation was not obtained. It was suggested that the cell membranes may act as ice nucleators. The nucleation temperature depended on the sample size (Fig. 3.6). Emulsified starch granules also contain catalytic sites of the required dimensions for promoting heterogeneous nucleation as cell membranes. The destruction of the granules increased the number of nucleation sites. A non emulsified starch suspension gave a value which was relatively close to the value obtained with emulsified starch. This was probably due to the fact that the suspension was at its "critical concentration", the concentration at which the amount of water around the granules is as low as it is in the emulsion.

We therefore conclude that the macromolecules and particularly food polymers encourage the nucleation of ice crystals. Thus they promote a thin ice texture in frozen products, which is generally looked for in the freezing process.

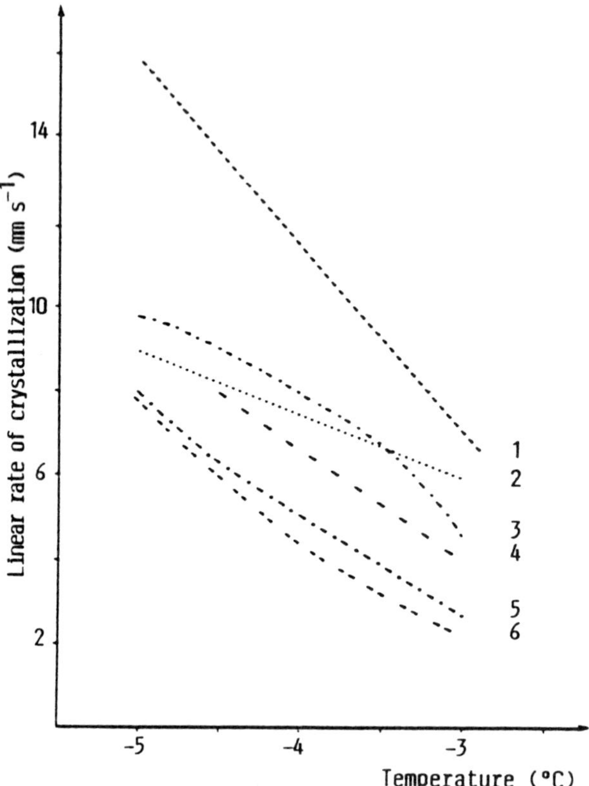

Fig. 3.7. Linear rate of ice crystallisation in suspensions of some hydrocolloids (concentration: 1 %).
1 Carrageenan, **2** CMC, **3** sodium Alginate (1.5 %), **4** tragacanth gum, **5** guar gum, **6** locust bean gum, redrawn from Budiaman and Fennema (1987).

Crystal Growth

A smooth crystalline texture in a frozen food is the consequence of a high nucleation rate associated with a limited rate for the crystal growth.

Considerable experimental data have been reported on the growth rate of ice crystals in undercooled water or solutions. Unfortunately the interpretation of these results is made difficult because the two factors which control the growth rate, namely the deposition kinetic and the dissipation of the latent heat of crystallisation, cannot be separated.

Experiments which provide information on the ice crystal growth can be divided in two classes (Hobbs 1974):

i water or solutions are contained in tubes, the measured growth rate being usually referred to as linear crystallisation velocity or propagation rate of the ice front;
ii the ice crystals grow freely in the bulk volume.

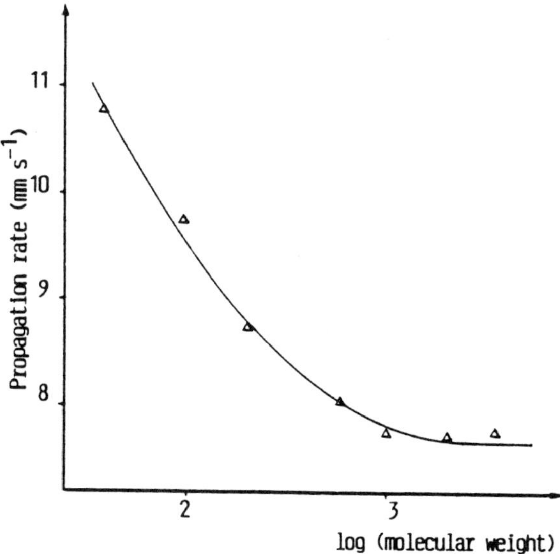

Fig. 3.8. Influence of the molecular weight of the polymer - Polyethylene glycol - on the ice propagation rate.
concentration = 2 %; undercooling = 5°C.

The first technique is more widely used; in these conditions the propagation rate of the ice front provides an overall estimation but when performed under standardised conditions it allows interesting comparative studies.

All experimental data reported on the growth of ice crystals in solutions indicate that all dissolved substances reduce the growth rate. A study of the parameters acting upon the propagation rate of the ice front shows the influence of the following:

The nature of the polymer In 1955 Lusena demonstrated that sugars and proteins could slow down the ice propagation more efficiently than mineral salts. Budiaman and Fennema (1987a) assessed the ability of various hydrocolloids to inhibit the rate of ice growth. As shown in Figure 3.7, their comparison for the same concentration demonstrates the large influence of the origin of the polymer, locust bean and guar gums being the most efficient.

The undercooling These measurements can only be performed in the range of undercooling temperature from 1°C to 6°C, spontaneous nucleation disturbing the experiments at lower temperature. The curve slopes (Fig. 3.7) indicate that all the polymers exhibit a large but different temperature dependence.

The concentration A large influence of the concentration on the ice propagation rate is observed. With solutions of low viscosity CMC (Blond 1988), it decreases from 13 mm/s to 3 mm/s when the concentration increases from 1% to 5%, for a constant undercooling temperature (ΔT = 5°C). Budiaman and Fennema (1987a) studied this effect with several hydrocolloids: the analysis of the linear crystallisation versus temperature also indicated a variation of the concentration effect with the origin of the polymer.

The molecular weight In aqueous solutions of PEG, the rate of propagation of the ice front decreases significantly with increasing molecular weight in the range 300-10 000 (Fig. 3.8). At higher molecular weights this effect is not observed.

The viscosity An increasing concentration produces, particularly with macromolecules, an increase of the viscosity. The experiments performed with mixtures of CMC of different molecular weights showed that the propagation rate of the ice front was weakly reduced even when the viscosity increased from 50 mPa.s to 1000 mPa.s (Blond 1988). Budiaman and Fennema (1987b) also studied the linear rate of water crystallisation at different viscosities but alteration of the viscosity was achieved by a change in the concentration of the solutes. They concluded that viscosity was not a good predictive criterion of hydrocolloids ability to inhibit crystallisation growth.

The viscosity is a macroscopic property of the solutions and the mobility of small solutes and of water is known to be as high in macromolecular systems as in pure water. The polymer concentration remains relatively low and it can thus be assumed that water diffuses freely in the medium.

The stuctural organisation The immobilisation of macromolecules by the formation of a structured network exerts a significant influence on the rate of propagation of the ice front. The importance of this effect may vary according to the nature and to the number of junctions in the network. Table 3.2 shows that the presence of a gel structure delays the ice propagation and that a further rate lowering is obtained by increasing the structure, for example with Ca^{++} bindings in low methoxyl pectin. Muhr and Blanshard (1986) also observed that the ice growth rate was altered more significantly in gelled alginate systems than in ungelled ones. It is obvious therefore that the ice growth is very sensitive to gel structure, even in sucrose solutions. On the contrary, Budiaman and Fennema (1987a) found no difference in the rate of ice crystallisation in fluid and gelled carrageenan.

It would be an interesting subject to go deeper into the relation between the rheological behaviour of the materials and the ice propagation. The structure effect seems to be associated with the elastic component of the mechanical properties of the systems. When the viscous component is large, ice propagation is fast; on the contrary, when the gel becomes elastic (matured gelatin, Table 3.2) a great lowering of the crystallisation velocity can be obtained. More experiments would be required to generalise this issue.

Table 3.2. Propagation rate of ice front in gel systems (Blond 1988)

	Propagation rate ($mm\ s^{-1}$)
Gelatin gel 2% matured	
0°C 1 h	6.52 A
24 h	6.04 A B
20°C 1 h	5.92 B
24 h	4.99 C
Low methoxyl pectin	
Solution: pectin 2%	11.35 E
Gel: pectin 2% + 3.38 m*M* Ca^{2+}	8.30 F
Gel: pectin 2% + 4.73 m*M* Ca^{2+}	7.23 G

Values marked with different letters are significantly different

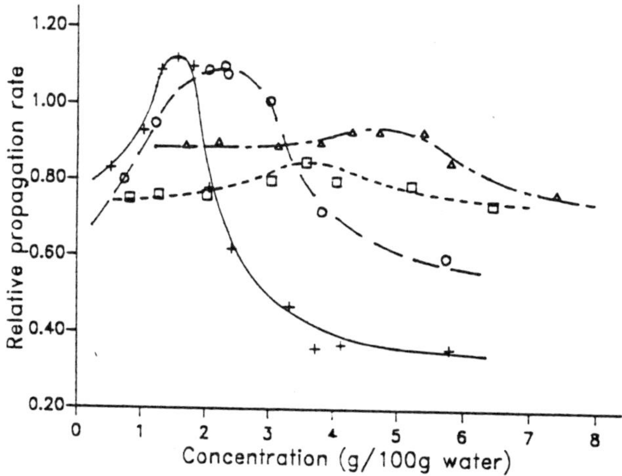

Fig. 3.9. Relative propagation rate of ice front as a function of concentration in crosslinked starch suspensions (rate in suspension/rate in pure water) (undercooling = 5°C) Crosslinking/anhydroglucose units ratio: + o, o 1/8000, 1/1500, Δ1/700.

It can be concluded that the decrease in the linear crystallisation velocity in macromolecular systems is not associated with a change in dynamic or thermodynamic properties of water. Crystallisation velocity is not limited by the diffusion of water but more probably by a mechanical hindrance exerted by the macromolecular solute. This essentially mechanical effect may be increased by increasing the concentration, but still more by increasing the length of the macromolecule. This latter increase being responsible for the entanglement of the chains. The immobilisation is amplified by gelification and the rigidity of the network may be expected to slow down the propagation of ice front.

The steric hindrance due to the polymer is confirmed by measurements with gelatinised starch granules (Fig. 3.9). A specific variation of the propagation rate can be observed as a function of granules concentration, the rate being higher than that in water when the granule concentration is close to the critical concentration. The freezing takes place out of the granules and the ice front propagation is very fast in the external water surrounding the granules. The external water volume being minimum at this concentration, the ice invasion rate is maximum decreasing when the crosslinking level increases. Above the critical concentration the main effect seems to be due to the rigidity of the granules. When the starch suspension is homogenised the propagation rate decreases regularly when the concentration increases. Similar results have been obtained with tomato cells (to be published).

When the tube technique is used, what is measured is the invasion rate of ice crystals and the real size of the crystals is not known. At a given temperature a low propagation rate allows the development of a higher number of nuclei and the resulting texture consists of small crystals.

Conclusion

We have attempted to demonstrate that the mechanism of ice formation is very complex even in simple polymer solutions. The frozen texture is the consequence of the nucleation and crystal growth processes which depend, on the one hand, on cooling rate and storage conditions, on the other hand, on the composition of the product including interactions between its components.

In addition, as the storage temperature is not always low enough, a recrystallisation process (or Ostwald ripening) can occur. In this process the ice crystal size changes, involving the growth of larger crystals at the expense of smaller ones, which reduces the advantages obtained by a fast freezing. This situation has been well exemplified with histological techniques applied to biological tissues.

It was expected that macromolecules could also limit the ice recrystallisation, but Cottrel *et al.*(1979) found that all the stabilisers they tested produced poor control of ice crystal growth and that the ice creams became icy over a 16 weeks storage at -35°C. Other works (Harper and Shoemaker 1983, Buyong and Fennema 1988) also showed that macromolecules had no stabilising influence on the crystal size during storage.

We have seen that, despite a high apparent viscosity, the water is very mobile in the solutions of macromolecules and that its diffusion is more limited by temperature than by solutes.

The practical objective might be to obtain a high nucleation (a large number of nuclei limiting the growth of each of them) and storage at the lowest realistic temperature in order to slow down the diffusion rate of water molecules. The ideal temperature is lower than the glass transition temperature. To slow down the diffusion rate of water molecules, a low storage temperature alone is sufficient. The effect of the mechanical hindrance given by a network structure on ripening during storage and the possibility to modify the product composition in order to increase the Tg of the freeze-concentrated phase could be studied. Levine and Slade (1989) have already indicated the efficiency of some polysaccharides to increase the Tg value.

References

Bigg EK (1953) as cited by Angels CA (1982) Supercooled water. In: Franks F(ed) Water : a comprehensive treatise, vol 7, Plenum Press, New York

Blond G (1985) Freezing in polymer-water systems and properties of water. In: Simatos D, Multon JL (eds) Properties of water in foods, NATO ASI Series, pp 531-542

Blond G (1986) Nucleation behaviour of water in macromolecular systems. Cryo-Lett 7:95-102

Blond G (1988) Velocity of linear crystallisation of ice in macromolecular systems.Cryobiology 25:61-66

Blond G (1989) Water-galactose system: supplemented state diagram and unfrozen water. Cryo-Lett 10:299-308

Budiaman ER, Fennema O (1987a) Linear rate of water crystallisation as influenced by temperature of hydrocolloid suspensions. J Dairy Sci 70:534-546

Budiaman ER, Fennema O (1987b) Linear rate of water crystallisation as influenced by viscosity of hydrocolloid suspensions. J Dairy Sci 70:547-554

Buyong N, Fennema O (1988) Amount and size of ice crystals in frozen samples as influenced by hydrocolloids. J Dairy Sci 71:2630-2639

Chen JY, Piva M, Labuza TP (1984) Evaluation of water binding capacity (WBC) of food fiber sources. J Food Sci 49:59-63

Clausse D, Jolivet C (1989) Aspects physico-chimiques de la cristallisation dans les systèmes dispersés : émulsions-brouillards. Société française des thermiciens Paris 10 May

Cottrell JIL, Pass G, Philips GO (1979) Assessment of polysaccharides as ice cream stabilisers. J Sci Food Agric 30:1085-1088

Franks F (1982) Apparent osmotic activities of water soluble polymers used as cryoprotectants. Cryo-Lett 3:115-120

Franks F, Asquith MH, Hammond C, Le B. Skaer H, Echlin P (1977) Polymeric preservation of biological ultrastructure. J Microsc 110:223-238

Franks F, Mathias SF, Parsonage P, Tang TB (1983) Differential scanning calorimetric study on ice nucleation in water and in aqueous solutions of hydroxyethyl starch. Thermochim Acta 61:195-202

Franks F, Mathias SF, Trafford K (1984) The nucleation of ice in undercooled water and aqueous poymer solutions. Colloid Surf 11:275-285

Franks F, Darlington J, Schenz T, Mathias SF, Slade L, Levine H (1987) Antifreeze activity of Antarctic fish glycoprotein and a synthetic polymer. Nature 325:146-147

Harper EK, Shoemaker CF (1983) Effect of locust bean gum and selected sweetening agents on ice recrystallisation rates. J Food Sci 48:1801-1803

Hoo AF, McLellan MR (1987) The contributing effect of apple pectin on the freezing point depression of apple juice concentrates. J Food Sci 52:372-377

Hobbs PV (1974) Ice physics Oxford University Press UK

Korber C, Scheiwe MW (1980) The cryoprotective properties of hydroxyethyl starch investigated by means of differential thermal analysis. Cryobiology 17:54-65

Kuhn W (1956) Uber die durch anomale kristallgestalt sowie durch limitierung der kristallgrosse bedingte gefrierpunkt- semiedrigung. Hev Chim Acta 39:1071-1086

Levine H, Slade L (1989) A food polymer science approach to the practice of cryostabilisation technology. Comments Agri & Food Chem 1:315-396

Lusena CV (1955) Ice propagation in systems of biological interest III. Effect of solutes on nucleation and growth of ice crystals. Arch Biochem. Biophys 55:277-284

Mackenzie AP (1975) The physicochemical environment during the freezing and thawing of biological materials. In: Duckworth RB (ed) Water relations of foods, Acad Press, London, pp 477-503

Mackenzie AP (1977) Non equilibrium freezing behaviour of aqueous systems, Phil Trans R Soc Lond B 278:167-189

Mackenzie AP, Rasmussen DH (1972) Interactions in water-PVP systems at low temperatures. In: Jellinek HHG (ed) Water structure at the water-polymer interfaces, Plenum Press, NY, pp 146-172

Mathias SH, Franks F, Trafford K (1984) Nucleation and growth in deeply undercooled erythrocytes. Cryobiology 21:123-132

Michelmore RW, Franks F (1982) Nucleation rates of ice in undercooled water and aqueous solutions of polyethylene glycol. Cryobiology 19:163-171

Muhr AH (1983) as cited by Muhr and Blanshard (1986)

Muhr AH, Blanshard JMV (1986) Effect of polysaccharide stabilisers on the rate of growth of ice. J Food Techn 21:683-710

Muhr AH, Blanshard JMV, Sheard SJ (1986) Effects of polysaccharide stabilisers on the nucleation of ice. J Food Techn 21:587-603

Parungo Fp, Wood J (1968) Freezing of aqueous solutions of macromolecules. J Atmos Sci 25:154-155

Pyenson H, Dahle CD (1938) Bound water and its relation to some dairy products. J Dairy Sci 21:169-185

Reid DS (1983) Fundamental physicochemical aspects of freezing. Food Technol 10:110-115

Rey DK, Labuza TP (1981) Characterisation of the effects of solutes on the water-binding and gel strength properties of Carrageenan. J Food Sci 46:786-789

Rey L (1964) Fundamental aspects of lyophilisation. In: Rey L (ed) Researches and development in freeze-drying, Hermann Paris pp 23-43

Simatos D, Blond G (1990) DSC studies and stability of frozen foods. In: Levine H and Slade L (eds) Water relationships in foods. Plenum Press Publ Co (in press)

Solms DJ, Rijke AM (1971) Anomalous freezing behavior of polymer gels and solutions. J Phys Chem 75:2623-2631

Voilley A, Blond G, Chevalier G (1989) Mesure et prediction de fortes activites de l'eau dans des systemes eau-polymères. Lebensm.-Wiss.u.-Technol 22:32-38

Willenger SA, Smith DE (1975) Effect of sweetener/stabiliser interactions on the viscosity and freezing point of ice cream mix. Michwissenschaft 41:766-769

Chapter 4

The Thermophysical Properties of Frozen Foods

C. A. Miles

'Concepts without factual content are empty; sense data without concepts are blind.... The understanding cannot see. The senses cannot think. By their union only can knowledge be produced.' (Immanuel Kant, 1724-1804, cited by Mackay, 1977).

Introduction

Perusal of any of the many published tables of the composition of foods (e.g. Paul and Southgate 1978) will show that most foods contain large quantities of water. When such foods are cooled below 0°C, a temperature is reached at which ice will form, and the highest temperature at which ice may exist in a food in thermal equilibrium is sometimes called the initial freezing point, or the transition temperature. Unlike pure water, all the water in food does not change to ice at this temperature. Only an infinitesimally small quantity of water crystallises at the transition temperature. As the temperature is reduced further, more water is frozen, the quantity increasing rapidly with decreasing temperature at first, and subsequently more slowly. The proportion of ice in any frozen food is therefore highly temperature dependent.

Since the physical properties of ice and water are very different, many of the properties of foods are governed by the ice content and the rate of change of ice content with temperature or pressure. For example, the specific enthalpy, density and thermal conductivity are dominated by the proportion of ice in the product, and the specific heat capacity, coefficient of cubical expansion at constant pressure (expansivity), and isothermal compressibility are dominated by the rate of change of the proportion of ice with temperature or pressure. We shall see later that, in aqueous foods, the rate of change of ice content with temperature at constant pressure varies in direct proportion to the rate of change of ice content with pressure at constant temperature.

The large latent heat of fusion of ice is particularly important to the refrigeration aspects of food freezing since it is responsible for the marked temperature dependence of the enthalpy and specific heat capacity of frozen foods. Understanding the equilibrium between ice and water in a frozen food thus forms the basis for understanding its thermophysical properties, and much of this paper is therefore concerned with the thermodynamics of ice/water systems. Only closed systems are considered and we will restrict ourselves to normal refrigeration temperatures, above -40°C. We will therefore not consider glass transitions which have been observed at lower temperatures in some frozen foods (Simatos *et al.* 1975).

The paper is written in sections of decreasing generality. The first section concerns the thermodynamics of ice/water systems in general, the second likens food freezing to the freezing of aqueous solutions, and the final section presents empirical equations for estimating the thermophysical properties of foods.

Thermodynamics of Food Freezing

Under conditions of constant pressure, P, and temperature, T, thermal equilibrium within a frozen food will be established at a particular ice content, α, the mass of ice divided by the total mass of the food. When P or T or both are changed, α also changes in order to re-establish thermal equilibrium, and we can consider the response of a frozen food to a change in pressure, volume or temperature to be composed of two terms: one representing the effect of a change in the proportion of ice, the other the response that would have occurred if the ice content were invariant.

For example, representing the entropy and volume of unit mass of food by S and V respectively, the following statements may be made:

$$\left(\frac{\partial V}{\partial T}\right)_P = \left(\frac{\partial V}{\partial T}\right)_{P,\alpha} + \left(\frac{\partial V}{\partial \alpha}\right)_{P,T}\left(\frac{\partial \alpha}{\partial T}\right)_P \qquad (1)$$

and

$$\left(\frac{\partial S}{\partial P}\right)_T = \left(\frac{\partial S}{\partial P}\right)_{T,\alpha} + \left(\frac{\partial S}{\partial \alpha}\right)_{P,T}\left(\frac{\partial \alpha}{\partial P}\right)_{T.} \qquad (2)$$

Equating the properties at constant ice content separately in Maxwell's equation

$$\left(\frac{\partial V}{\partial T}\right)_P = -\left(\frac{\partial S}{\partial P}\right)_T$$

yields

$$\left(\frac{\partial V}{\partial \alpha}\right)_{P,T}\left(\frac{\partial \alpha}{\partial T}\right)_P = -\left(\frac{\partial S}{\partial \alpha}\right)_{P,T}\left(\frac{\partial \alpha}{\partial P}\right)_T$$

i.e.

$$-\Delta V\left(\frac{\partial \alpha}{\partial T}\right)_P = \frac{L}{T}\left(\frac{\partial \alpha}{\partial P}\right)_T \qquad (3)$$

where

$$\Delta V = -\left(\frac{\partial V}{\partial \alpha}\right)_{P,T} \tag{4}$$

and

$$L = -T\left(\frac{\partial S}{\partial \alpha}\right)_{P,T}. \tag{5}$$

That is to say, L and ΔV are respectively the heat absorbed by the food and the change in the volume of the food, when unit mass of ice within it melts under conditions of constant temperature and pressure. To calculate numerical values of some of the thermophysical properties of foodstuffs, L may be equated with the latent heat of fusion of ice, and ΔV with the change in volume which occurs when unit mass of ice melts. Strictly this is an approximation, since, in foodstuffs, the ice which separates on freezing is abstracted from solutions of increasing concentration and not from pure water. There will be small contributions to L and ΔV caused by concentrating the solutes, since the partial molal entropy and volume of the solutes will, in general, vary with concentration. These effects disappear entirely only at infinite dilution.

When Equation 3 is inserted in the mathematical identity

$$\left(\frac{\partial P}{\partial T}\right)_{\alpha}\left(\frac{\partial \alpha}{\partial P}\right)_{T}\left(\frac{\partial T}{\partial \alpha}\right)_{P} = -1$$

it gives an important result, analogous to Clapeyron's equation, establishing the slope of the equilibrium P, T curves for frozen foods:

$$\left(\frac{\partial P}{\partial T}\right)_{\alpha} = \frac{L}{T\Delta V} \tag{6}$$

The right hand side of Equation 6 is the same as Clapeyron's equation for a first order transition (see any thermodynamic text book, e.g. Pippard 1961), but the left hand side is a partial differential defining a set of P, T loci at set ice contents. Clapeyron's equation defines a single P, T curve for the equilibrium between the two phases.

The specific heat capacity at constant pressure, $C_{P,}$ may be written in two parts:

$$C_P = T\left(\frac{\partial S}{\partial T}\right)_P = T\left(\frac{\partial S}{\partial T}\right)_{P,\alpha} + T\left(\frac{\partial S}{\partial \alpha}\right)_{P,T}\left(\frac{\partial \alpha}{\partial T}\right)_P$$

i.e. from (5)

$$C_P = C_{P,\alpha} - L\left(\frac{\partial \alpha}{\partial T}\right)_P \tag{7}$$

where

$$C_{P,\alpha} = T\left(\frac{\partial S}{\partial T}\right)_{P,\alpha} \tag{8}$$

is the specific heat capacity at constant ice content, sometimes referred to as the sensible specific heat capacity. Similarly, dividing Equation 1 by V, (=1/ρ), and substituting Equation 4, the coefficient of cubical expansion, θ, can be written:

$$\theta = \frac{1}{V}\left(\frac{\partial V}{\partial T}\right)_P = \theta_\alpha - \rho\Delta V\left(\frac{\partial \alpha}{\partial T}\right)_P \tag{9}$$

where

$$\theta_\alpha = \frac{1}{V}\left(\frac{\partial V}{\partial T}\right)_{P,\alpha} \tag{10}$$

Analogous expressions for the isothermal compressibility, K_T, are:

$$K_T = -\frac{1}{V}\left(\frac{\partial V}{\partial P}\right)_T = -\frac{1}{V}\left(\frac{\partial V}{\partial P}\right)_{T,\alpha} - \frac{1}{V}\left(\frac{\partial V}{\partial \alpha}\right)_{P,T}\left(\frac{\partial \alpha}{\partial P}\right)_T$$

i.e.

$$K_T = K_{T,\alpha} + \Delta V\rho\left(\frac{\partial \alpha}{\partial P}\right)_T \tag{11}$$

where

$$K_{T,\alpha} = -\frac{1}{V}\left(\frac{\partial V}{\partial P}\right)_{T,\alpha} \tag{12}$$

Substituting Equation 3 in 11 yields:

$$K_T = K_{T,\alpha} - \frac{T\rho\Delta V^2}{L}\left(\frac{\partial \alpha}{\partial T}\right)_P \tag{13}$$

Putting

$$\delta C_p = C_p - C_{p,\alpha} \tag{14}$$

$$\delta\theta = \theta - \theta_\alpha \tag{15}$$

and

$$\delta K_T = K_T - K_{T,\alpha} \tag{16}$$

yields, from Equations 7, 8 and 12, the following inter-relationships:

$$\frac{\delta C_P}{L} = \frac{\delta\theta}{\rho \Delta V} = \frac{\delta K_T L}{T\rho\Delta V^2} \tag{17}$$

The specific heat capacity at constant volume, C_v, and the adiabatic compressibility, K_s, can be determined from Equations 7, 8 and 13 using the general thermodynamic relations (see for example Pippard 1961):

$$C_v = C_p - T\frac{\theta^2}{\rho K_T} \tag{18}$$

and

$$K_s = \frac{C_v}{C_p} K_T \tag{19}$$

Finally, thermodynamics text books (e.g. Pippard 1961) show that the temperature change induced by reversible adiabatic compression may be calculated from:

$$\left(\frac{\partial T}{\partial P}\right)_S = \frac{T\theta}{\rho C_p} \tag{20}$$

which for a frozen food, using Equations 7 and 9, may be written as:

$$\left(\frac{\partial T}{\partial P}\right)_S = \frac{T\Delta V}{L}\left(1 - \frac{C_{p,\alpha}}{C_p} + \frac{L\theta_\alpha}{\rho C_p \Delta V}\right) \tag{21}$$

For high water content foods, a first approximation, that near the transition temperature:

$$\frac{C_{p,\alpha}}{C_p} - \frac{L\theta_\alpha}{\rho C_p \Delta V} << 1$$

yields

$$\left(\frac{\partial T}{\partial P}\right)_S \approx \frac{T\Delta V}{L} \tag{22}$$

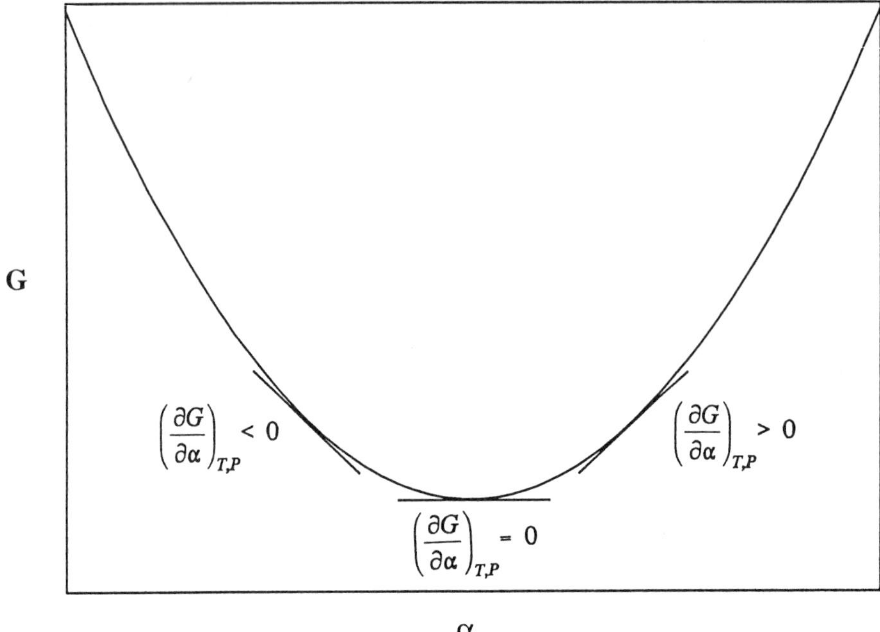

Fig. 4.1. How the Gibbs free energy at constant pressure and temperature shows a minimum at the equilibrium ice content.

That is, provided the approximation holds, near the transition temperature all aqueous foods yield approximately the same temperature change on reversible adiabatic compression, and this change depends only on the thermodynamic properties of the pure water/ice transition.

There is an alternative way of deriving the above equations. The mathematical analysis follows closely that of Lamb's (1965) analysis of thermal mechanisms for the absorption of ultrasound in liquids, but the meaning of the symbols is different. Consider an aqueous food containing ice. We require three independent variables to define the state of the closed system. For example we could choose temperature, pressure and ice content. By analogy with the affinity of chemical reactions (e.g. Katchalsky and Curran 1967) we will define the affinity, A, of the ice/water transition, a measure of the departure of the system from equilibrium, as:

$$A = -\left(\frac{\partial G}{\partial \alpha}\right)_{P,T}$$

where G is the Gibbs free energy. At a given temperature and pressure the system has a minimum G at a given value of α, the equilibrium value (see Fig. 4.1).
Thus

$$-dG = SdT - VdP + Ad\alpha \qquad (23)$$

and when the system is in equilibrium A = 0.

The change in the affinity for the ice/water transition due to a change in temperature, pressure and α may be written:

$$dA = \left(\frac{\partial A}{\partial T}\right)_{P,\alpha} dT + \left(\frac{\partial A}{\partial P}\right)_{T,\alpha} dP + \left(\frac{\partial A}{\partial \alpha}\right)_{P,T} d\alpha \tag{24}$$

But by Maxwell's relations:

$$\left(\frac{\partial A}{\partial T}\right)_{P,\alpha} = \left(\frac{\partial S}{\partial \alpha}\right)_{P,T} = -\frac{L}{T} \tag{25}$$

and

$$\left(\frac{\partial A}{\partial P}\right)_{T,\alpha} = -\left(\frac{\partial V}{\partial \alpha}\right)_{P,T} = \Delta V \tag{26}$$

So

$$dA = -\frac{L}{T} dT + \Delta V dP + \left(\frac{\partial A}{\partial \alpha}\right)_{P,T} d\alpha \tag{27}$$

When the changes are made so slowly that α is allowed to reach an equilibrium value at each instant, A = 0 and dA = 0
and

$$\left(\frac{\partial A}{\partial \alpha}\right)_{P,T} d\alpha = -\Delta V dp + \frac{L}{T} dT \tag{28}$$

Therefore

$$\left(\frac{\partial P}{\partial T}\right)_{\alpha} = \frac{L}{T\Delta V} \tag{29}$$

This is the same result as Equation 6.
In fact all the equations previously established by equilibrium thermodynamics may be deduced from this approach. For example,

$$dS = \left(\frac{\partial S}{\partial T}\right)_{P,\alpha} dT + \left(\frac{\partial S}{\partial P}\right)_{T,\alpha} dP + \left(\frac{\partial S}{\partial \alpha}\right)_{P,T} d\alpha \tag{30}$$

Therefore

$$C_p = T\left(\frac{\partial S}{\partial T}\right)_P = C_{P,\alpha} - L\left(\frac{\partial \alpha}{\partial T}\right)_P \tag{31}$$

in agreement with Equation 7.

We know that A=0 at equilibrium and when dα, the change in α from its equilibrium value, is negative, A is positive and when dα is positive, A is negative. Therefore the product Adα < 0 and the change in G due to any displacement of the reaction away from the equilibrium position is positive. The affinity, A, may therefore be considered as the driving force for the ice/water transition.

Provided the displacement does not deviate too far from equilibrium, the simplest assumption is that

$$\frac{d\alpha}{d\tau} = mA$$

where m is a constant and τ is time.
But at constant P and T

$$dA = \left(\frac{\partial A}{\partial \alpha}\right)_{P,T} d\alpha$$

Therefore, by integration, since A = 0 at equilibrium:

$$A = \left(\frac{\partial A}{\partial \alpha}\right)_{P,T} (\alpha - \bar{\alpha})$$

near equilibrium, where $\bar{\alpha}$ is the equilibrium value.
Hence

$$\frac{d\alpha}{d\tau} = m\left(\frac{\partial A}{\partial \alpha}\right)_{P,T} (\alpha - \bar{\alpha})$$

or

$$\frac{d\alpha}{d\tau} = -\frac{(\alpha - \bar{\alpha})}{\tau_0}$$

i.e.

$$\alpha - \bar{\alpha} = (\alpha_0 - \bar{\alpha})\exp\left(-\frac{\tau}{\tau_0}\right)$$

where τ_0, the characteristic time for the relaxation, is given by:

$$\frac{1}{\tau_0} = -m\left(\frac{\partial A}{\partial \alpha}\right)_{P,T}$$

In fact, when Kent and Jason (1975) studied the time dependence of microwave attenuation in frozen cod, stored isothermally following freezing, they found it decayed with two time constants, τ_1 and τ_2.

Experimental Data

There are several lines of evidence which support the above analyses.

1 Morley (1986) measured $(\partial T/\partial P)_s$ in frozen beef at the initial freezing point and found it equal to (-0.70 ± 0.02) x 10^{-7} K Pa^{-1}. This may be compared with the theoretical prediction of Equation 21. At -1°C, L = 332 kJ/kg and ΔV=-0.903 x $10^{-4} m^3$/kg (data taken from Miles and Morley, 1978).

$C_{P,\alpha}/C_P$ can be estimated from Riedel's (1978) data to be 0.026 at the initial freezing point, and $L\theta_\alpha/\rho C_p \Delta V$ is of the order of -0.003 (see Miles and Morley 1978). Hence Equation 21 yields a theoretical prediction for $(\partial T/\partial P)_s$ of -0.72 x 10^{-7} K Pa^{-1}, in agreement with Morley's experiment.

2 Morley (1986) measured the coefficient of cubical expansion in frozen beef muscle and Riedel (1978) provides data from its specific heat capacity. These two sets of measurements may be compared with the predictions of Equation 17, re-written as:

$$\theta - \theta_\alpha = \rho \frac{\Delta V}{L}\left(C_p - C_{p,\alpha}\right)$$

Over the temperature range -1°C to -10°C, where most of the variation in θ and C_p is apparent, $\rho\ \Delta V/L$ varies from -2.94 x 10^{-7} to -2.86 x 10^{-7} and a mean value of -2.90 x 10^{-7} kg/J may be assumed. As a first approximation θ_α may be taken as constant in Equation 23, and a plot of θ against C_p-$C_{p,\alpha}$ should therefore be approximately linear. The good agreement between theory and experiment is shown in Figure 4.2, where the line has the theoretical slope, -2.90 x 10^{-7} kg/J, and an intercept, 7.8 x 10^{-4} K^{-1}.

3 As predicted, there is a discontinuity in $(\partial T/\partial P)_s$ at the initial freezing point. In the presence of ice $(\partial T/\partial P)_s$ is large and negative; in its absence, it is small and positive. The temperature of this discontinuity is therefore the initial freezing point. Using an apparatus similar to that described by Morley (1986), the initial freezing point of beef muscle was determined over a range of pressures and the results are plotted in Figure 4.3. The slope of the least squares line was(-8.13 ± 0.69) x 10^{-8} K Pa^{-1}, not significantly different from that predicted by Equation 6, -7.4 x 10^{-8} K Pa^{-1}. The intercept, the initial freezing point at atmospheric

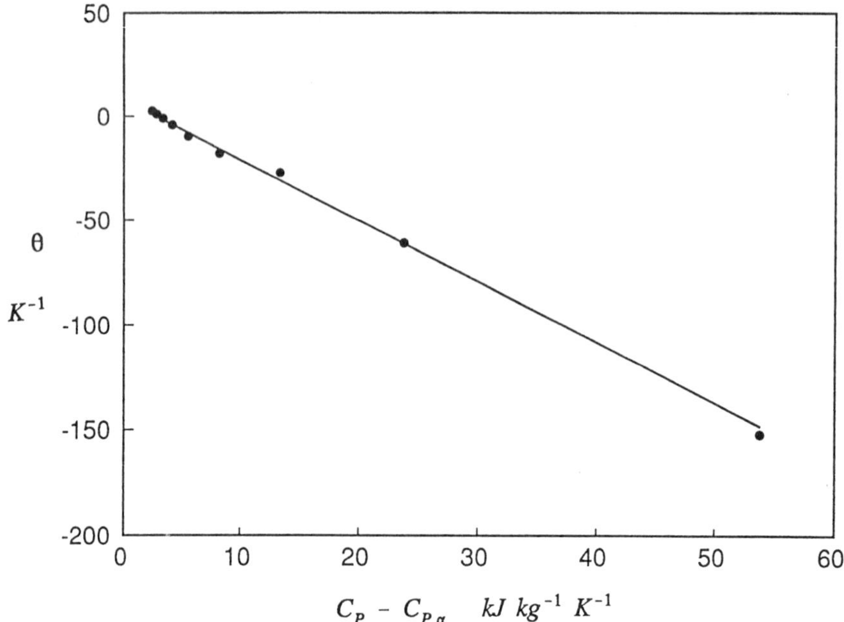

Fig. 4.2. Relation between the expansivity and specific heat capacity of frozen beef muscle. Data of Morley (1986) (expansivity) and Riedel (1978) (specific heat capacity). The text presents an equation (Equation 17) predicting a linear relation of slope $\rho\Delta V/L$, the slope of the line drawn in this figure.

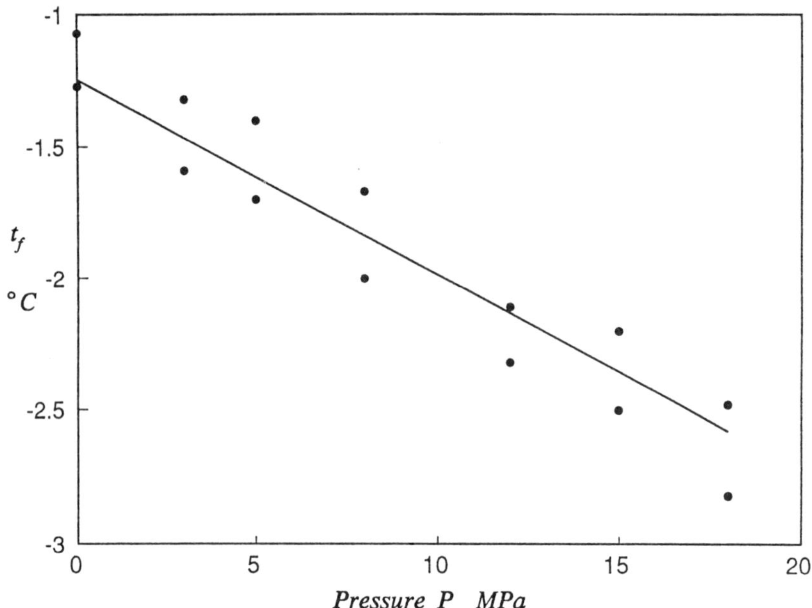

Fig. 4.3. Relation between the initial freezing point and environmental pressure in post-rigor muscle. Data of Cornick (personal communication). The line is the theoretical line (Equation 6), of slope -7.4×10^{-8} K Pa^{-1}

pressure, was approximately 0.2 °C lower than that usually quoted for beef. This may be due to an artifact in this experiment. When the meat was loaded into the pressure cell and the system 'bled' to remove all air some exudate was lost and the residual tissue might therefore be unrepresentative of intact muscle.

Freezing Point Depression in Foods

The above analyses do not explain the observation that the initial freezing point of foods is lower than the freezing point of pure water, and that the ice content increases rapidly at first with decreasing temperature and then gradually more slowly. To explain these facts we need a further hypothesis. In this section we will analyse the freezing of foods in terms of the freezing point depression caused by dissolved solutes, as proposed by many previous authors (e.g. Heldman 1974; Schwartzberg 1976).

At equilibrium, the chemical potentials of the ice and aqueous phases in a frozen food are equal. If we assume that the ice phase is pure ice and denote its chemical potential by $\mu^*_w(s)$, that of the liquid phase by $\mu_w(l)$ and that of pure water by $\mu^*_w(l)$, we may write, at equilibrium:

$$\mu_w^*(s) = \mu_w(l) = \mu_w^*(l) + RT\ln(a_w) \tag{32}$$

(See any text book on Physical Chemistry, e.g. Atkins 1978).

Here a_w is the water activity of the aqueous phase and R is the gas constant. The difference between the Gibbs free energies of pure water and ice is given by:

$$\Delta G = \mu_w^*(l) - \mu_w^*(S) \tag{33}$$

Therefore

$$\ln(a_w) = -\frac{\Delta G(T)}{RT} \tag{34}$$

Following Atkins (1978), when $a_w = 1$, $T = T^*$ and

$$\ln(1) = -\frac{\Delta G(T^*)}{RT^*}$$

Therefore

$$\ln(a_w) = -\left(\frac{\Delta G(T)}{RT} - \frac{\Delta G(T^*)}{RT^*}\right)$$

Putting $\Delta G = \Delta H - T\Delta S$ (H enthalpy; S entropy) yields the well-known result:

$$\ln(a_w) = \frac{\Delta H(T - T^*)}{RTT^*} \tag{35}$$

i.e.

$$a_w = \exp\frac{\Delta H(T - T^*)}{RTT^*} \tag{36}$$

For a solution obeying Raoult's law, a_w is the mole fraction of water in the aqueous phase. This may be written in terms of X_w = total mass of water/mass of food, X_b = mass of unfreezable water/mass of food, M_w = molecular mass of water, M_s = effective molecular mass of non-aqueous solids, and α = mass of ice/mass of food.

$$a_w = \frac{X_w - X_b - \alpha}{X_w - X_b - \alpha + \frac{M_w}{M_s} X_s} \tag{37}$$

where

$$X_s = 1 - X_w$$

Inserting 37 in 36 and using the fact that $\alpha = 0$ when $T = T_f$, the initial freezing point, yields:

$$\alpha = (X_w - X_b)\left(1 - \frac{\exp\frac{\Delta H(T - T^*)}{RTT^*}\left(1 - \exp\frac{\Delta H(T_f - T^*)}{RT_f T^*}\right)}{\exp\frac{\Delta H(T_f - T^*)}{RT_f T^*}\left(1 - \exp\frac{\Delta H(T - T^*)}{RTT^*}\right)}\right) \tag{38}$$

Note that for small excursions from the initial freezing point and for $T_f \approx T^*$, Equation 38 approximates to

$$\alpha = (X_w - X_b)\left(1 - \frac{(T_f - T^*)}{(T - T^*)}\right)$$

i.e.

$$\alpha = (X_w - X_b)\left(1 - \frac{t_f}{t}\right) \tag{39}$$

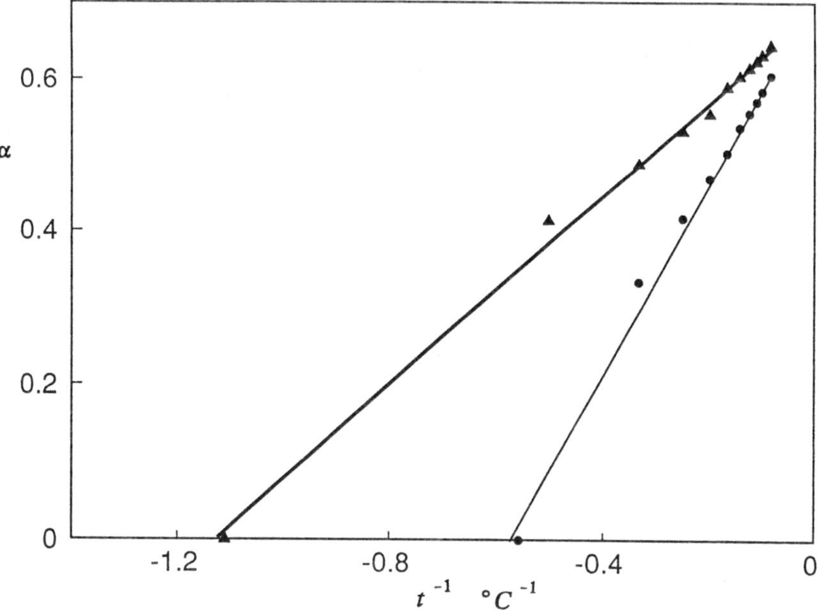

Fig. 4.4. Linear relations between the proportion of ice in salted beef (thin line) and unsalted beef (thick line) and the reciprocal of the temperature in degrees Celsius. The linear relation is predicted by Equation 39 of the text. Data of Morley, personal communication.

where t_f and t are respectively the initial freezing point and the temperature of the food in degrees Celsius. This equation was found empirically to fit experimental data for the ice content of beef, lamb, pork and cod (Miles 1974) and further evidence for it is given in Figure 4.4.

The equation has not been explicitly quoted in American work but has been widely used in Europe and is the basis of the COSTHERM program for determining the thermal properties of food (Miles *et al.* 1983). It may however be deduced from one of Schwartzberg's (1976) equations:

$$C_p = A + \frac{B}{\left(T^* - T\right)^2} \tag{40}$$

(Schwartzberg's Equation 27)

From Equation 7 we can see that

$$A = C_{p,\alpha}$$

and

$$\frac{B}{(T^* - T)^2} = -L\frac{\partial\alpha}{\partial T}$$

Put

$$T^* - T = -t$$

Hence

$$d\alpha = \frac{-B}{Lt^2}dt$$

which on integration, and inserting $\alpha = 0$ at $t = t_f$, yields:

$$\alpha = \frac{-B}{Lt_f}\left(1 - \frac{t_f}{t}\right)$$

Schwartzberg uses a different notation to that employed here. In his notation:

$$B = \left(n_{w_o}(1+b) - b\right)\Delta H_o(T_o - T_1)$$

To convert to our notation, put $T_o = T^*, T_1 = T_f, \Delta H_o = L, n_{w_o} = X_w$,

and $b = X_b/1 - X_w$ to yield:

$$B = -(X_w - X_b)Lt_f$$

Hence

$$\alpha = (X_w - X_b)\left(1 - \frac{t_f}{t}\right)$$

This is the same as Equation 39.

Schwartzberg states, in his notation, that

$$A = C_m - \left\{n_{w_o}(1+b) - b\right\}\Delta C_p$$

which converts, in our notation to:

$$A = C_{p_m} - (X_w - X_b)\left(C_{p_w} - C_{Pice}\right)$$

This is indeed an approximation for $C_{p,\alpha}$. A better approximation is given later (Equation 42). Schwartzberg's Equation 27 is therefore consistent with our Equations 39 and 7.

Equations for the ice content of foods based on the behaviour of aqueous solutions predict a discontinuity in the slope of the ice content/temperature curve at the initial freezing point and this implies that there is also a discontinuity in the specific heat capacity. Riedel (1978), however, reports that his measurements of the specific heat capacity are continuous in partially dehydrated foods at low hydrations. Foods having a high water content show a sharp maximum which resembles a discontinuity. Whereas the freezing point mechanism may therefore be predominantly that of aqueous solutions at high hydrations, at low liquid water content (high solute concentrations), other mechanisms may become increasingly important. For example Haynes (1968) has discussed freezing point depression in porous substances caused by surface tension effects, and non-equilibrium effects, such as those reported by Kent and Jason (1975), may be increasingly important at low hydrations and temperatures because of the long time constants.

Empirical Equations

Knowing the relation between ice content and temperature (e.g. Equation 39), the specific heat of a frozen food can be calculated from Equation 7. But there are some parameters in these equations that have to be estimated. In Equation 39 the 'bound' water content may be estimated by various methods, see for example Chen (1985), but a simple equation, used in the COSTHERM program for estimating the thermal properties of foods (Miles *et al.* 1983) is:

$$X_b = 0.3X_p + 0.1X_c \tag{41}$$

where X_p is the mass fraction of protein and X_c the mass fraction of carbohydrate. Use of Equation 39 with Equation 41 allows estimation of the ice content of an aqueous food of known composition at any temperature and fixed pressure.

$C_{p,\alpha}$, the specific heat capacity at constant ice content, may be estimated from:

$$C_{p,\alpha} = C_{p_m} - \left(C_{p_w} - C_{p_{ice}}\right)\alpha \tag{42}$$

where the specific heat capacity of the unfrozen product, C_{pm} is given by:

$$C_{p_m} = X_w C_{p_w} = X_p C_{p_p} + X_f C_{p_f} + X_c C_{p_c} + X_{min} C_{p\,min} \tag{43}$$

i.e. simplifying the notation

$$C_{pm} = \Sigma X_i C_i$$

and numerical estimates for the specific heat capacities of protein, fat, carbohydrate and minerals (respectively C_{pp}, C_{pf}, C_{pc} and C_{pmin}) are given in Table 4.1.

Table 4.1. Estimates of the physical properties of food components

Component	Density kg m^{-3}	Specific heat capacity J kg^{-1}K^{-1}	Thermal conductivity W m^{-1}K^{-1}
Water[a] t> 4°C	1002.7 - .275 t	4218	.554 + .00165 t
Water[a] t< 4°C	999.8 + .130 t		
Ice	917	1961	780/T - .615
Protein	1380	1900	.2
Fat	930	1900	.18
Carbohydrate	1550	1220	.25
Minerals	2400	840	.25

t temperature °C; T temperature K
[a] a single function for the density of water is given by Kell (1967)

The specific enthalpy of the food (here arbitrarily set to zero at t=0) may be obtained by integration:

$$h = \int_o^t C_p dT \tag{44}$$

The density of the unfrozen product in the absence of gaseous inclusions is given by

$$\rho_{\varepsilon=o} = \frac{1}{\Sigma \dfrac{X_i}{\rho_i}} \tag{45}$$

and that of a degassed frozen product by:

$$\frac{1}{\rho} = \frac{1}{\rho_{\varepsilon=o}} - \alpha\left(\frac{1}{\rho_w} - \frac{1}{\rho_{ice}}\right) \tag{46}$$

For a product containing air (porosity ε), the density ρ_ε is given by

$$\rho_\varepsilon = (1-\varepsilon)\rho_{\varepsilon-o} \tag{47}$$

The isothermal compressibility of a product containing no air may be estimated from Equation 11 using:

$$K_{T,\alpha} = \frac{\rho}{\rho_m} K_{Tm} - \alpha\rho\left(\frac{K_{Tw}}{\rho_w} - \frac{K_{Tice}}{\rho_{ice}}\right) \tag{48}$$

and K_{Tm}, the isothermal compressibility of the unfrozen product is given by:

$$K_{Tm} = \Sigma \varepsilon_i K_{Ti} \tag{49}$$

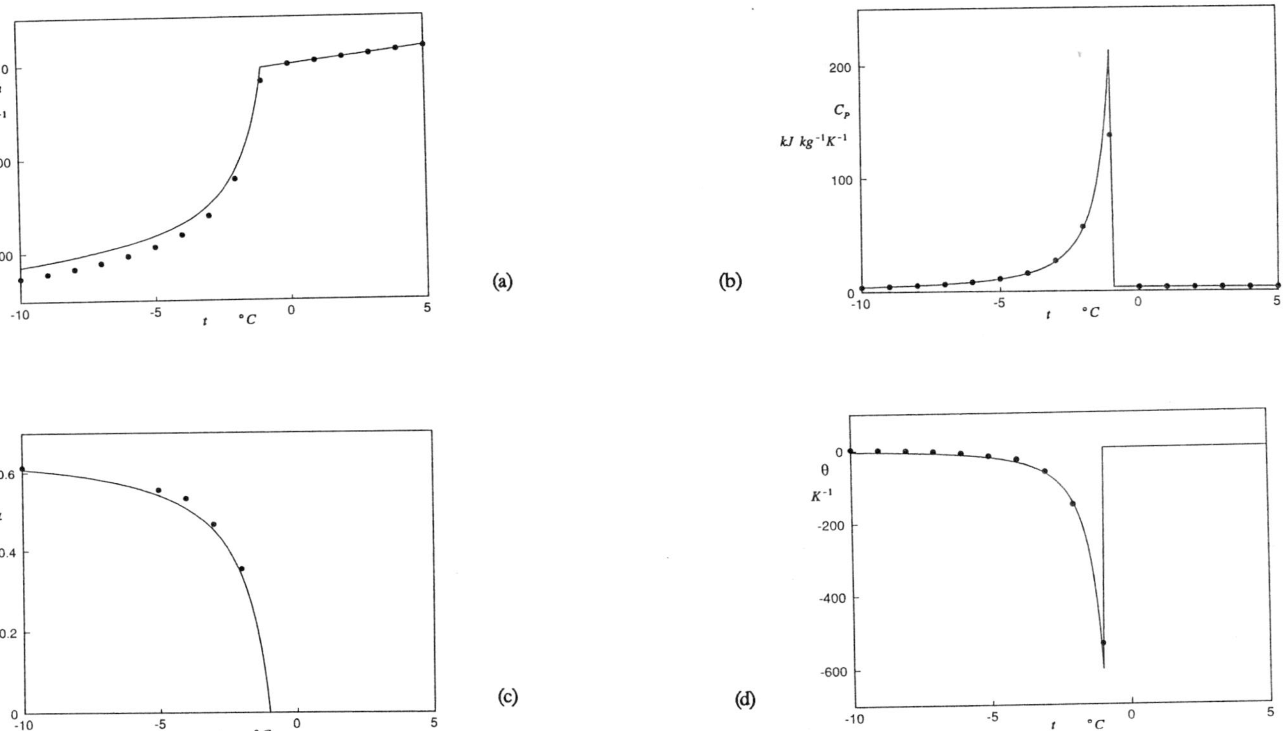

Fig. 4.5. Estimates of the thermophysical properties of beef muscle (lines) compared with experimental data. Computations made on the basis of the following composition: 74% water, 21.4% protein, 3.9% fat, 0.7% minerals. **a** Enthalpy; Equation 43; data of Riedel 1977 **b** Specific heat capacity; Equation 7; data of Riedel 1977 **c** Ice content; Equation 39; data of Riedel 1977**d** Expansivity; Equation 9; data of Morley 1986.

where ε_i is the volume fraction of the ith component. In the unfrozen product K_T, K_S and K_{Tm} might be estimated from the speed of ultrasound, c, since $K_T \approx K_S \approx (\rho c^2)^{-1}$ (see Miles 1974b).

The expansivity of a frozen product containing no air may be estimated in the region of the initial freezing point using Equation 9, but is likely to be highly dependent upon hydration and temperature (see Morley and Miles 1988), so the equation is most useful when

$$\left|\rho\Delta V\left(\frac{\partial\alpha}{\partial T}\right)_P\right| >> |\theta_\alpha|$$

and the second term in Equation 9 dominates.

Having estimated C_p, and K_T, the specific heat capacity at constant volume, C_v, and the adiabatic compressibility, K_s, can be estimated from Equations 18 and 19 respectively.

The accuracy of the above equations may be judged from Figure 4.5, which superimposes experimental values, on numerical estimates for beef.

The form of the temperature dependence of the thermal conductivity of a frozen product has been estimated from the following equation:

$$\lambda = \frac{\rho\lambda_m}{\rho_m} - \alpha\rho\left(\frac{\lambda_w}{\rho_w} - \frac{\lambda_{ice}}{\rho_{ice}}\right) \tag{50}$$

and λ_m, the thermal conductivity of the unfrozen product, is given by:

$$\lambda = \Sigma\varepsilon_i\lambda_\iota \tag{51}$$

Estimates for the thermal conductivities of food components are given in Table 4.1. Once λ, C_p and ρ have been estimated, the thermal diffusivity, a, may be obtained from the equation:

$$a = \frac{\lambda}{\rho c} \tag{52}$$

However the agreement between experimental and estimated values for λ and a is not good. Equations 50 and 51 were derived on the basis of a parallel model (Miles *et al.* 1983). Better estimates may be obtained by introducing an empirical geometry factor, f, specific to the particular product type, and Equation 50 becomes (Morley, personal communication):

$$\lambda \approx \lambda_m - \alpha f(\lambda_w - \lambda_{ice}) \tag{53}$$

For fibrous foods, such as meat and fish, f is approximately 0.61 for heat flow perpendicular to the fibres and 0.77 for parallel heat flow (Fig. 4.6). In comminuted products f is intermediate and a value of 0.7 may be used.

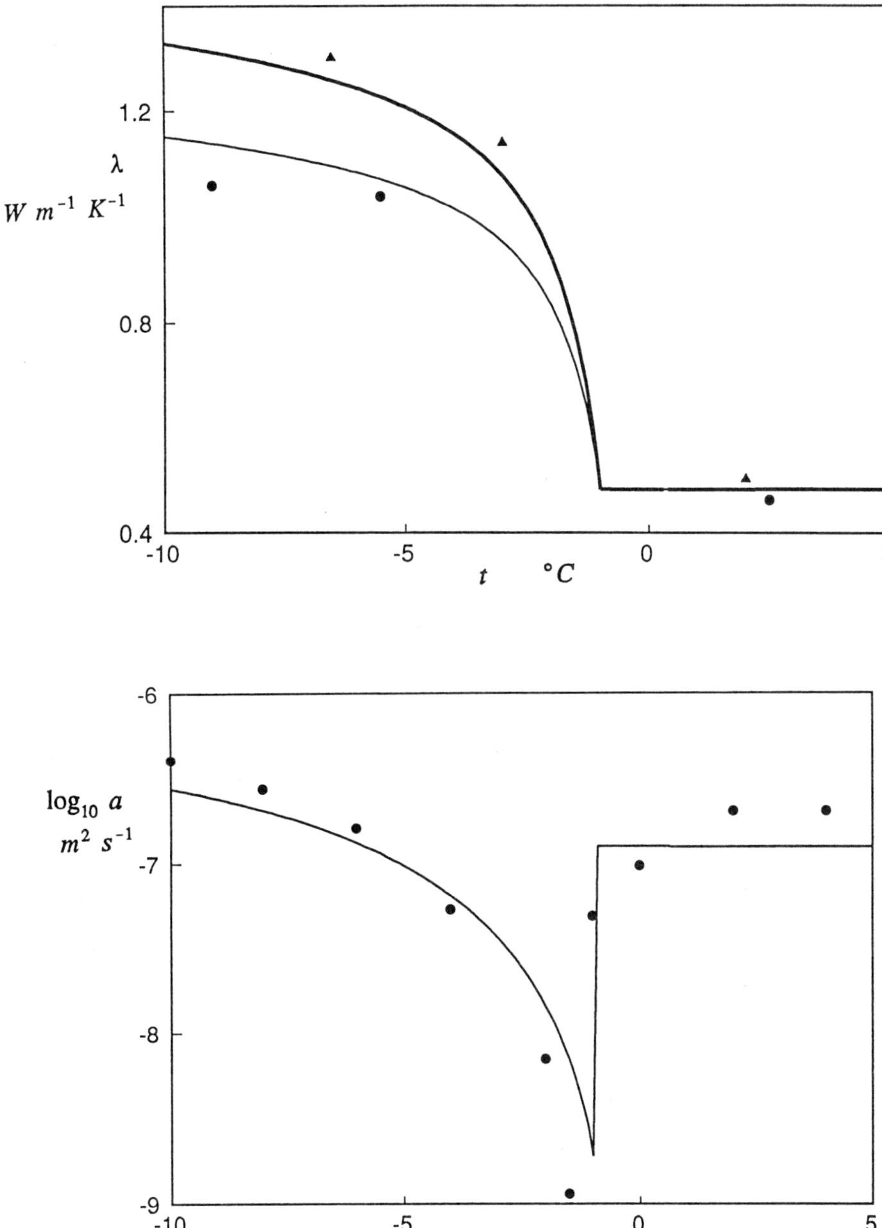

Fig. 4.6. Estimates of the thermal conductivity and thermal diffusivity of muscle. **a** Thermal conductivity of beef (composition as Figure 4.5): thick line heat flow parallel to fibres (Equation 53, f = 0.77) data of Lentz (1961); thin line heat flow perpendicular to fibres (Equation 53; f=0.61) data of Lentz (1961) **b** Thermal diffusivity perpendicular to the fibres of cod muscle (estimated composition 79% water, 20% protein, 0.6% fat, 0.4% minerals): line (Equation 52, f = 0.61) data of Nesvadba 1982.

Conclusions

1 The specific heat capacities at constant pressure and constant volume, the coefficient of thermal expansion, and the isothermal and adiabatic compressibilities of frozen foods are all inter-related via Equations 17, 18 and 19.
2 Frozen foods with a high water content and containing no gas are more compressible near the initial freezing point than the fresh product.
3 At constant pressure the proportion of ice in a frozen food decreases with temperature at a rate that is proportional to its rate of change with pressure at constant temperature (Equation 3).
4 The temperature fall induced by reversible adiabatic compression of a high water content frozen food, very close to its initial freezing point, depends primarily on the thermodynamic properties of the pure ice/water transition and is approximately the same for all high water content foods.
5 The change in temperature required to maintain equilibrium in a frozen food undergoing a change in hydrostatic pressure is given by Equation 6.
6 At constant pressure, the ice content of aqueous frozen foods changes with temperature approximately according to Equation 39.
7 Empirical equations can be used to predict the following thermophysical properties: density, coefficient of cubical expansion, specific heat capacity at constant pressure and constant volume, specific enthalpy, isothermal and adiabatic compressibility, thermal conductivity and thermal diffusivity.

Acknowledgements

I wish to record my thanks to Mr MJ Morley for helpful discussion on the thermal properties of foods over many years and to Mr GAJ Fursey for help in the preparation of this manuscript.

References

Atkins PW (1978) Physical Chemistry, Chapter 8. Oxford University Press, Oxford

Chen CS (1985a) Thermodynamic analysis of the freezing and thawing of foods: enthalpy and apparent specific heat. Journal of Food Science 50:1158-1162

Chen CS (1985b) Thermodynamic analysis of the freezing and thawing of foods: ice content and Mollier diagram. Journal of Food Science 50:1163-1166

Haynes JM (1968) Thermodynamics of freezing in porous solids. In: Hawthorne J, Rolfe EJ (eds) Low temperature biology of foodstuffs. Pergamon Press, Oxford pp 79-104

Heldman DR (1974) Predicting the relationship between unfrozen water fraction and temperature during freezing using freezing point depression. Trans ASAE 17:63

Katchalsky A, Curran PF (1967) Non equilibrium thermodynamics in biophysics. Harvard University Press, Cambridge, Massachusetts

Kell GS (1967) Precise representation of volume properties of water at one atmosphere. Journal of Chemical and Engineering data 12:67-69

Kent M, Jason AC (1975) Dielectic properties of foods in relation to interactions between water and substrate. In: Duckworth RB (ed) Water relations of foods. Academic Press, New York pp 211-231

Lamb J (1965) Thermal relaxation in liquids. In: Mason WP (ed) Physical Acoustics vol IIA, Academic Press pp 203-280

Lentz CP (1961) Thermal conductivity of meats, fats, gelatin gels and ice. Food Technology, Champaign 15:243-247

Mackay AL (1977) The harvest of a quiet eye. The Institute of Physics, Bristol and London p 84

Miles CA (1974a) The ice content of frozen meat and its measurement using ultrasonic wares. In: Meat Freezing - How and Why? Meat Research Institute, Langford, Bristol pp 15.1-15.7

Miles CA (1974b) Relationships between some physical properties of frozen meat. In: Meat Freezing - How and Why? Meat Research Institute, Langford, Bristol pp 16.1-16.10

Miles CA, Morley MJ (1978) The effect of hydrostatic pressure on the physical properties of frozen foods. J Physics D: Applied Physics 11:201-207

Miles CA, Van Beek G, Veerkamp CH (1983) Calculation of thermophysical properties of foods. In: Jowitt R, Esher F, Hallstrom B, Meffert HFTh, Spiess WEL, Vos G (eds) Physical properties of food. Applied Science, London

Morley MJ (1986) Derivation of physical properties of muscle tissue from adiabatic pressure-induced temperature measurements. J Food Technology 21:269-277

Morley MJ, Miles CA (1988) Influence of the degree of hydration on the thermal expansion of muscle tissue. International Journal of Food Science and Technology 23:177-184

Nesvadba P (1982) A new transient method for the measurement of temperature dependent thermal diffusivity. J Physics D: Applied Physics 15:725-738

Paul AA, Southgate DAT (1978) McCance and Widdowsons The Composition of Foods, 4th edn HMSO

Pippard AB (1961) Elements of classical thermodynamics. Cambridge University Press, Cambridge

Riedel L (1978) Eine Formel zur Berechnung der Enthalphie fettarmer Lebensmittel in Abhangigkeit von Wassergehalt und Temperatur. Chem Mikrobiol Technol Lebensm 5:129-133

Schwartzberg HG (1976) Effective heat capacities for the freezing and thawing of foods. Journal of Food Science 41:153

Simatos D, Faure M, Bonjour E, Couach M (1975) Differential thermal analysis and differential scanning calorimetry in the study of water in foods. In: Duckworth RB (ed) Water relations of foods, Academic Press, London

Chapter 5

Ice Crystal Growth in Idealised Freezing Systems

W. B. Bald

Introduction

It is a generally accepted postulation that the quality and texture of frozen food is improved by minimising the size of ice crystals formed during the freezing process. This postulation also extends into the storage period of frozen foods. According to Arbuckle (1986), for example, the mean size of ice crystals in ice cream should be less than about 45μm to 55 μm to satisfy the customers' tastes.

Many research workers in the field of Cryobiology have shown experimentally that ice crystal size in any frozen system is reduced by increasing the cooling rate. One of the pioneers in this work was Luyet who, along with his co-workers, defined in a qualitative way the type of ice crystals which formed in different solutions. Figure 5.1 is a typical example of this work which shows the decreasing size and increasing number of dendritic ice crystals which form in a 50% polyvinylpyrrolidone (PVP) solution as the freezing temperature decreases. The decreasing freezing temperature is synonymous with an increase in the applied cooling rate.

Real food systems are structurally much more complex than aqueous solutions because of the cell walls, muscle tissue, etc., but a quantitative analysis of idealised freezing systems can greatly assist in the understanding of actual frozen food products.

It is not clear at present how theoretical models derived for aqueous solutions will apply to real biological specimens and foodstuffs. In real systems the unfrozen water in combination with other unfreezable solids encompass the ice crystals in an agglomeration whose viscosity changes rapidly with temperature. These viscosity changes could affect significantly ice crystal growth during freezing and subsequent storage.

The following discussion is the author's views on how changes in ice crystal size can be predicted for idealised aqueous systems. Considerable experimental work needs to be carried out to confirm the analysis and to extend the approach to real food products.

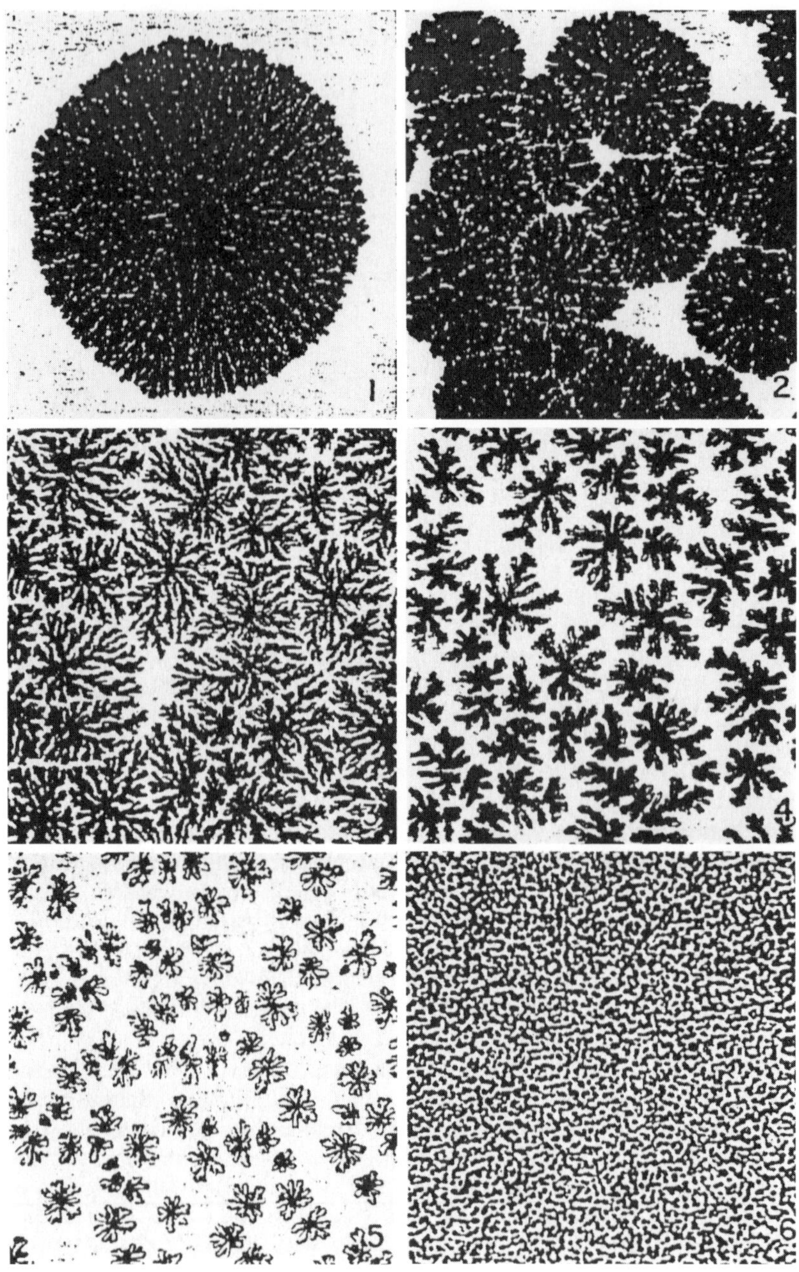

Fig. 5.1. Size and number of irregular dendrites of ice developed in a given area, in the freezing of a 50% solution of polyvinylpyrrolidone. Freezing temperatures: Photos 1 to 6: -40°F (-40°C), -49°F (-45°C), -58°F (-50°C), -67°F (-55°C), -76°F (-60°C), and -85°F (-65°C) respectively Magnification: 100X. From Luyet and Rapatz (1958)

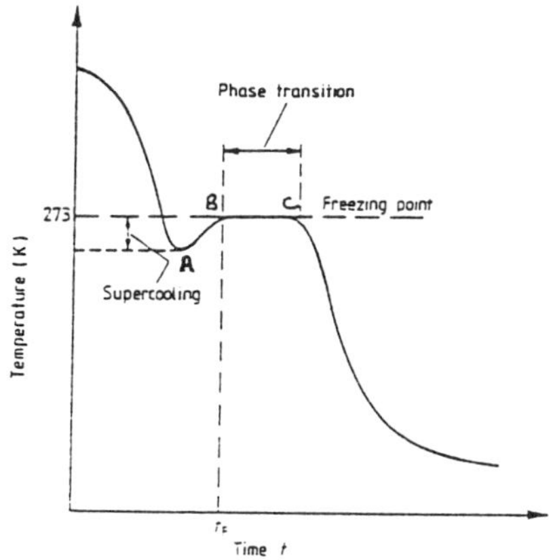

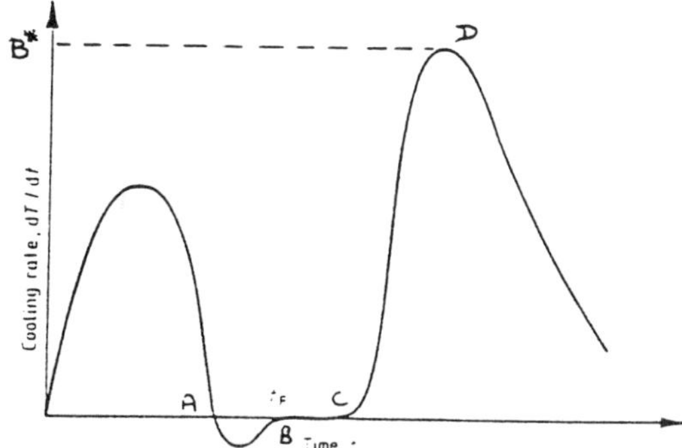

Fig. 5.2a. Typical temperature-time curve in the small localised region of ice crystal formation in an aqueous system.
Fig. 5.2b. The slope of Fig 5.2a or localised cooling rate variation in a freezing aqueous system.

Critical Cooling Rate B*

We start our discussion by considering the temperature changes at some very small location within a biological sample or food product undergoing continuous global cooling. The temperature will change within this small region in the manner shown in Figure 5.2a and the corresponding slope of this curve or cooling rate varies as shown in Figure 5.2b. In practice a small micro-thermocouple located within the sample would not detect the change A-B-C but would indicate a more or less continuous curve. (See for example the measurements of Meryman 1966).

Point A signifies the nucleation temperature where a critical nucleus of radius r* forms in the liquid phase producing a localised warm up or negative cooling rate as shown. Ice crystal growth will then propagate from this nucleus during the small time interval A-B-C.

At point C the cooling rate and accompanying heat flux in this small region will rapidly increase to point D to match the surrounding global heat flux. The cooling rate at point D in Figure. 5.2b is the critical cooling rate B* which controls the extent or limit for ice crystal growth. This value corresponds with the global cooling rate at the freezing front B_f when it reaches the small region under discussion.

It should be noted that this global cooling rate B_f is the maximum which occurs at a specific region within a frozen product. Diller (1990a) has recently suggested that the cooling rate which controls ice crystal size is some averaged value over a specific temperature change at any location. It would seem, however, that this approach will not produce an explicit relationship between the cooling rate and ice crystal size.

Ice Crystal Size and Cooling Rate

Many of the early attempts to provide a quantitative relationship between mean crystal size $\bar{s}$ and cooling rate at the freezing front B_f have been derived for the rapid cooling of metals. According to Jones (1986) the dendritic cell size $\bar{s}$ is related to the cooling rate by the empirical expression

$$\bar{s} = mB_f^{-n} \qquad (1)$$

where m and n are constants and typically $0.25 \leq n \leq 0.5$.

Although the general form of Equation 1 should be applicable to the freezing of foodstuffs, the constants could be significantly different because of the difference in Biot modulus Bi and Stefan number St between liquid metals and biological systems. (See Bald 1988).

A possible model for determining the relationship between cooling rate and ice crystal size is to treat the critical nucleus of radius r* as a continuous point heat source as suggested by Riehle and Hoechli (1973). This approach, however, cannot produce an explicit solution for the ice crystal size without pre-supposing values for the extent of the heat source.

A more plausible approach is to apply the solution for the spherical phase growth problem given by Frank (1950). Using this solution produces values for the heat flux and cooling rate at a given ice crystal size which are much greater than the observed values. This is due to the growth of a complete sphere rather than an ice crystal made up of dendritic spears as shown in Figure 5.1.

In 1986 the author proposed a model for predicting the mean size of ice crystals in an idealised freezing system by representing the dendritic growth by the model system shown in Figure 5.3. The general relationship between ice crystal size and critical cooling rate was derived as

$$B_f = B^* = \frac{\pi U \rho_1 L}{128\, K_1 (\bar{s}/2r^* - 1)^2} \frac{dX}{dt} \{1 + [1 + 32(\bar{s}/2r^* - 1)]^{1/2}\}^2 \qquad (2)$$

where dX/dt is the velocity of the moving freezing front as it propagates into the food product. This velocity will vary depending on the method used to freeze a specific product.

For plate freezers the propagating velocity is given by (Bald 1987).

$$\frac{dX}{dt} = 2\lambda^2 \alpha_1 / X \qquad (3)$$

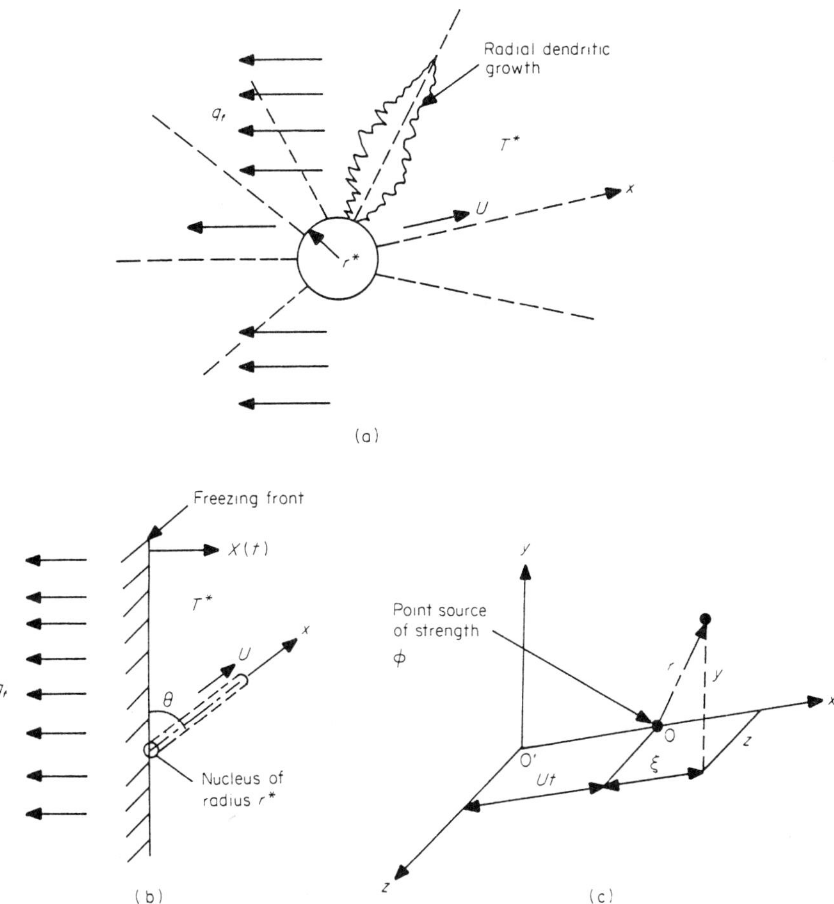

Fig. 5.3. Theoretical model of a growing ice crystal. (**a**) Dendritic growth along random radial spears at velocity U in a uniform heat flux field q_f. (**b**) Assumed cylindrical heat source propagating at velocity U in the x direction at some angle θ to the freezing front in a uniform temperature field T* and heat flux q_f (**c**) Co-ordinate system for a propagating point heat source of strength Φ at point o moving in the x direction at constant velocity U.

whereas for freezing in a tunnel with constant surface heat transfer coefficient the value is

$$\frac{dX}{dt} = \left(\frac{K_1 B^*}{\rho_1 L}\right)^{1/2} \tag{4}$$

Figure 5.4 shows a typical plot of the decreasing cooling rate B* as the ice crystal size increases for two different assumed values for the dendritic ice growth rate U. These theoretical curves representing freezing in a tunnel cooler were obtained using

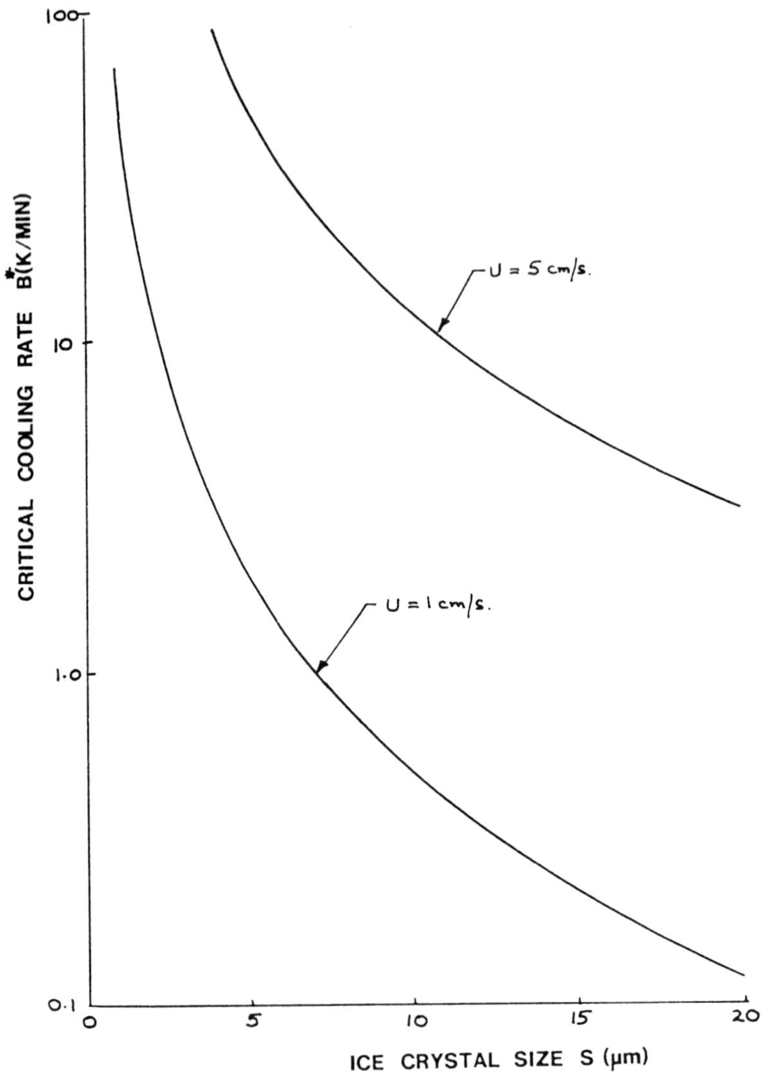

Fig. 5.4. Typical curves showing the variation of cooling rate with ice crystal size calculated from the theoretical model of Equations 2 or 6. Numerical values given in the text.

Equations 2 and 4 with the assumed numerical values L = 250 J/G, ρ_1 = 0.98 g/cc, K_1 = 0.019 W/cmK and r* = 5 nm.

In the case of food freezing $\overline{s}/2r^* \gg 1$ in which case Equation 2 simplifies to

$$B^* = \frac{\pi U \rho_1 L}{4K_1(\overline{s}/2r^*)} \frac{dX}{dt} \tag{5}$$

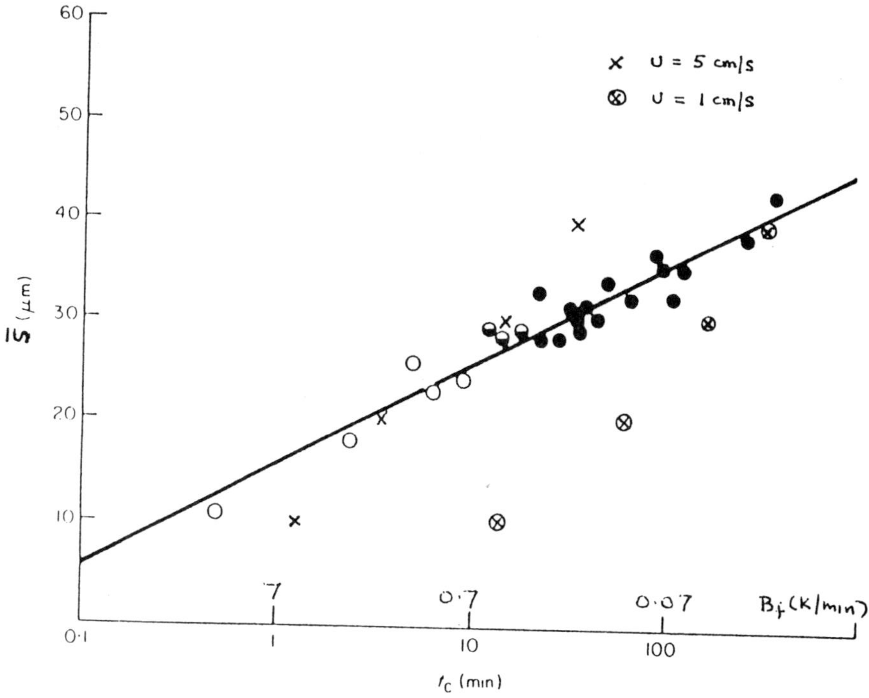

Figure 5.5. Comparison between theoretical model Equation 2 and the measured ice crystal sizes for frozen beef by Bevilacqua *et al.* (1979). ⊗U = 5 cm/sec ⊗ U = 1 cm/sec. K_1 = 0.0152 W/cmK, c = 4 J/gK, L = 183.4 J/g, T* = 272 K, ρ_1 = 0.95 g/cm^3, r* = 5nm.

Substituting Equation 4 into 5 gives for tunnel freezing

$$\bar{s} = \frac{\pi U r^*}{2} \sqrt{\frac{\rho_1 L}{K_1}} \quad B^{*-1/2} \tag{6}$$

which is of the same general form as Equation 1 where

$$m = \frac{\pi U r^*}{2} \sqrt{\frac{\rho_1 L}{K_1}} \tag{7}$$

A comparison between the theoretical values obtained from Equation 2 and the measured average ice crystal diameters in frozen beef by Bevilacqua *et al.* (1979) is shown in Figure 5.5. This comparison suggests that the dendritic ice growth rate U may be a function of the cooling rate and not constant as assumed in the derivation of Equation 2.

Cooling Rates in Real Food Products

The foregoing section has discussed the inter-dependence between the ice crystal size and cooling rate. To estimate the variation in ice crystal size throughout a food product therefore, values for the global cooling rate at specific locations must be known.

An accurate assessment of the cooling rate variation in both time and space, within a specific food product during freezing, can only be obtained using computational numerical procedures. However, the author has proposed approximate pseudo steady state solutions for the cooling rate at the freezing front as it propagates into the product (Bald 1987).

Plate Freezers

In a plate freezer maintained at a constant plate temperature T_p the cooling rate at any distance X from the plate is given by

$$B_f = \frac{4\,(T_F - T_P)\,\lambda^3 \propto_1}{X^2 \sqrt{\pi}\ \mathrm{erf}\,(\lambda)} \exp(-\lambda^2) \tag{8}$$

where the parameter λ is given approximately by the equation

$$\lambda^2 = \frac{c_1\,(T_F - T_P)}{2[L + c_2\,(T_S - T_F)]} \tag{9}$$

Diller (1990b) has recently tabulated more accurate values for the parameter λ.

Equation 8 shows that the cooling rate B_f decreases as the square of the distance from the cooling plates. However, in the freezing of a slab between a double plate freezer the cooling rate at the centre will rapidly increase above the value predicted by Equation 8. This is the so called "centre line effect" in freezing problems which cannot be predicted by Equation 8.

If the cooling rate distribution predicted by Equation 8 is compared with the critical cooling rate values of Equation 2 required to produce an ice crystal size $\overline{s}$, small ice crystal sizes would be predicted at the slab surface increasing in size within the product except at the centre itself.

Tunnel Freezers

The pseudo steady state solution to the problem of food freezing in an air cooled or liquid nitrogen cooled tunnel assumes that both the mean surface heat transfer coefficient $\overline{h}$ and the coolant temperature T_∞ are constant.

Bald (1987) has proposed solutions to this problem for flat slab shaped products, cylindrical shaped products and spherical products. The derived cooling rate values at the freezing front are as follows

for a slab,

$$B_f = \frac{\overline{h}^2\,(T_F - T_\infty)^2}{\rho_1 L K_1 (1 + \overline{h}X/K_1)^2} \tag{10}$$

for a cylinder, of radius R,

$$B_f = \frac{\bar{h}^2 (T_F - T_\infty)^2 R^2}{\rho_1 LK_1 (R - X)^2 \{1 + (\bar{h}R/K_1) \ln [R/(R - X)]\}^2} \tag{11}$$

for a sphere, of radius R,

$$B_f = \frac{\bar{h}^2 (T_F - T_\infty)^2 R^4}{\rho_1 LK_1 (R - X)^4 \{1 + (\bar{h}R/K_1) [X/(R - X)]\}^2} \tag{12}$$

At the centre of a slab the "centre line effect" already discussed for plate freezers will occur which is not predicted by Equation 10. Equations 11 and 12, however, predict an infinite cooling rate as $X \rightarrow R$ which is in agreement with the experimental measurements of Meryman (1966).

Also in agreement with Meryman's observations are a predicted minimum in the cooling rate distributions of Equations 11 and 12 provided the Biot modulus $Bi>1$ in the case of the cylinder, and $Bi>2$ for the sphere. These theoretical conclusions could have a significant effect on the ice crystal size distribution throughout different shaped food products and on the ultimate quality.

In a cylindrical shaped frozen product for example, the minimum cooling rate will occur where $X = R[1 - \exp(K_1/\bar{h}R - 1)]$ provided $\bar{h}R/K_1>1$. This would be the region where the largest ice crystals would be likely to occur.

Ice Crystal Size Distribution in a Frozen Product

The foregoing sections have discussed the relationships between the cooling rate at some specified spatial location within a frozen product and the mean ice crystal size $\bar{s}$ in that region. At any specific location there will be a statistical distribution of ice crystal sizes around this mean. This could have a significant effect on recrystallisation during storage and subsequently on the ultimate food quality and texture.

Smith and Schwartzberg (1985) have measured the ice crystal size distributions in sugar solutions and shown that the number of ice crystals of a given diameter follow a typical normal or Gaussian distribution.

Calvelo (1981) showed how recrystallisation during the frozen storage of beef increased both the mean ice crystal size and the standard deviation. (See Fig 5.6)

Martino and Zaritzky (1988) produced histograms of the mean equivalent ice crystal diameter for frozen beef which again illustrated the normal form of the size distribution. This work also confirmed how the mean ice crystal size and the standard deviation increased with increasing storage time.

Assuming, therefore, a normal distribution of ice crystal sizes around the mean $\bar{s}$ obtained from Equation 2, (or Equation 6 in the case of a tunnel freezer), the number of ice crystals n(s) of a given size s is given by

$$n(s) = \frac{1}{\sigma\sqrt{2\pi}} \exp - \left[\frac{(s - \bar{s})^2}{2\sigma^2}\right] \tag{13}$$

where σ is the standard deviation of the observed distribution.

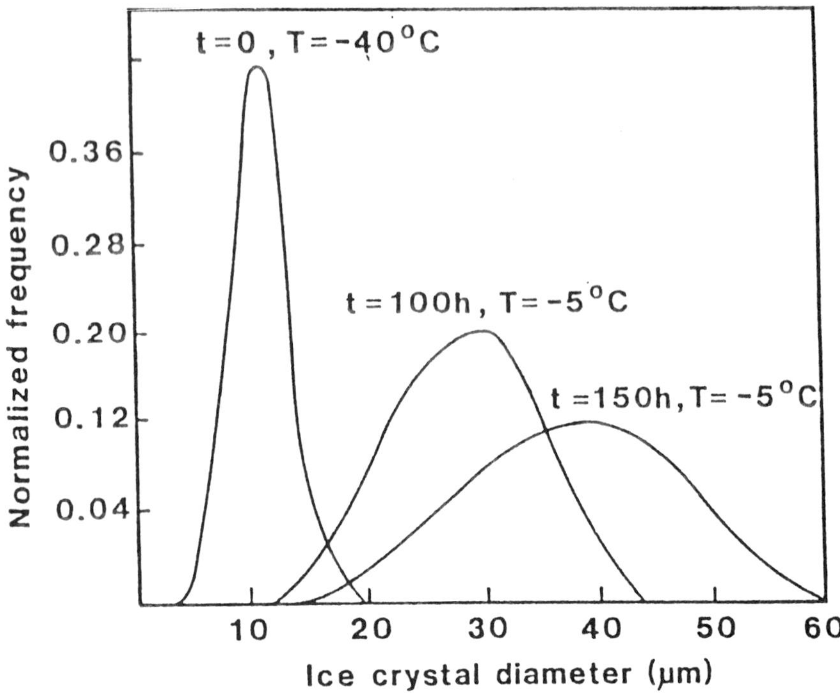

Figure 5.6. Effects of recrystallisation on the frozen storage of beef. From Calvelo (1981)

Thus Equation 13 can be used to define the number and distribution of ice crystals in a frozen product in the region of the specified cooling rate. For example by combining Equations 2 and 11 to find $\bar{s}$ the size distribution of ice crystals at some radius (R-X) in a cylindrical shaped product can be estimated.

If a global or overall ice crystal size distribution for any frozen product is required then the weighted mean cooling-rate expressions derived by Bald (1987) should be used in combination with Equation 2 to find $\bar{s}$.

Frozen Food Storage

The work of Calvelo (1981), Smith and Schwartzburg (1985), Martino and Zaritzky (1988) and many others have shown that during frozen food storage at a constant temperature recrystallisation takes place and specific ice crystals increase in size. This is thought to be caused by the migration of the unfrozen water surrounding each ice crystal towards the region of lower concentration.

Martino and Zaritzky (1988) have recently proposed the following semi-empirical relationship for the growth of the mean ice crystal size during storage.

$$\ln\left[\frac{\bar{S}_{max} - \bar{S}_o}{\bar{S}_{max} - \bar{S}}\right] + \frac{\bar{S}_o - \bar{S}}{\bar{S}_{max}} = \frac{Ct}{\bar{S}_{max}^{\,2}} \qquad (14)$$

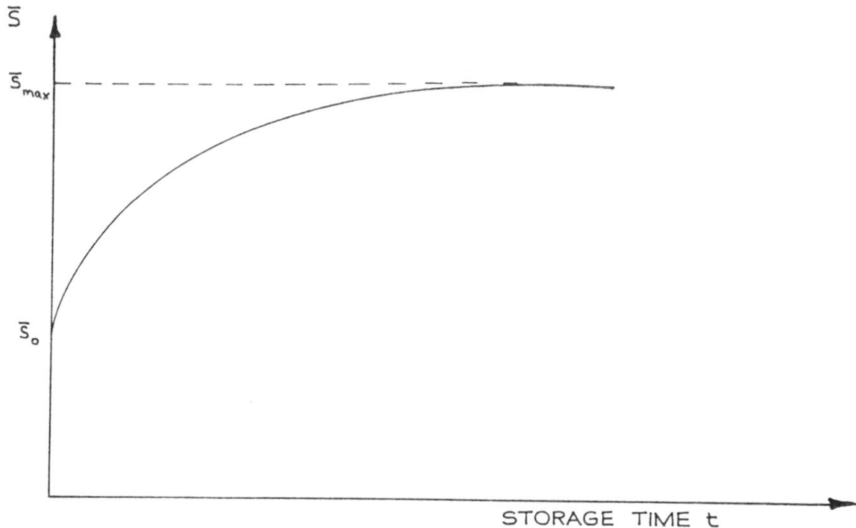

Fig. 5.7. Typical plot showing variation in mean ice crystal size $\bar{s}$ with storage time t.

C is a kinetic recrystallisation constant which varies for different frozen products and for different storage temperatures. Martino and Zaritzky (1988) experimentally observed the following recrystallisation constants for frozen beef tissue at different temperatures as shown in Table 5.1.

Table 5.1. Recrystallisation constants for frozen beef

Storage temperature (°C)	C (μm^2/day)
-5	339.4 ± 47.17
-10	200.34 ± 20.5
-15	142.45 ± 7.17
-20	107.39 ± 16.4

Figure 5.7 shows a typical plot of the variation in mean ice crystal size $\bar{s}$ with storage time t for a specific storage temperature calculated using Equation 14. $\bar{s}_o$ is the mean diameter of ice crystal immediately after freezing at time zero and $\bar{s}_{max}$ is the asymptotic value for large storage times.

A limitation on the value for $\bar{s}_{max}$, could, however, be caused by coalescence of adjacent ice crystals.

An estimate of the probable value for $\bar{s}_{max}$ can be derived from probability theory as originally applied to nucleate boiling of liquids by Gaertner (1963) and Bald (1973).

As the ice crystals grow during storage the upper limit $\bar{s}_{max}$ will occur when adjacent crystals or nearest neighbours coalesce. The most probably nearest neighbour distance $z = \bar{s}_{max}$ derived from statistical analysis is found to be

$$\bar{s}_{max} = \frac{1}{\sqrt{2\pi N''}} \tag{15}$$

where N" is the average population density of ice crystals (i.e. number of observed ice crystals per unit area) at the specified location in the frozen product. Statistical analysis also shows that 62% of the ice crystals at a specific location should satisfy Equation 15.

Using Equations 14 and 15 together with the initial mean ice crystal size $\bar{s}_0$ obtained from Equation 2 should enable the characteristics of the ice formation in frozen food storage to be ascertained.

Finally, it is worth noting, that the actual ice crystals observed in a particular frozen product, by either optical or electron micoscopy, will present a complex two-dimensional image. The equivalent diameter Se of each individual ice crystal is normally defined as

$$Se = \sqrt{4A/\pi} \tag{16}$$

where A is the measured area of the image.

An estimate of the total volume of ice in a frozen product is then

$$V_i = \frac{\pi}{6} \int_v \left[\sum_0^\infty s_e^3 n(s_e)\right] dV \tag{17}$$

where $n(s_e)$ is estimated using Equation 13.

Conclusions

The foregoing discussion is primarily a theoretical assessment of the problem of ice formation in frozen foodstuffs. Extensive experimental analysis and observations are necessary to check the validity of these Equations.

The validity of Equation 2, for example, depends on the assumption that U>>dX/dt and therefore the formation of an ice crystal will take place at a specific cooling rate near the freezing temperature of the water.

The pseudo steady state solutions for the cooling rate (Equations 10 - 12) require verification by using methods similar to Meryman (1966) but with closer control of the surface boundary conditions.

Extensive image analysis in combination with both optical and electron microscopy is necessary to verify the statistical distribution of ice crystal sizes within a frozen product and also the standard deviation caused by different cooling techniques. These methods will also have to be used to ascertain the kinetic recrystallisation constant and validity of Equation 14 for the long term storage of frozen foodstuffs.

A less empirical approach than Equation 14 for determining ice crystal growth during storage will require solutions to the phase growth problem as outlined by Frank 1950. This growth is controlled by concentration gradients in the surrounding medium caused by the release of unfrozen water in the amorphous matrix due to sudden changes in viscosity.

An additional important factor arises when considering the relationship between ice crystal formation and cooling rate in any product. An upper limit on cooling rate exists above which mechanical thermal stresses destroy the texture of the product and render it unacceptable to the customer.

Notation

A	measured area of ice crystal image
B^*	critical cooling rate
B_f	cooling rate at the propagating freezing front
C	Kinetic recrystallisation constant
c_1	specific heat of frozen phase
c_2	specific heat of unfrozen phase
h	mean surface heat transfer coefficient
K_1	thermal conductivity of frozen phase
L	latent heat of fusion
m	constant in Equation 1
n	constant in Equation 1, number
N''	number of observed ice crystals per unit area
r^*	critical radius of ice nucleus
R	radius of cylindrical or spherical product
$\overline{s}$	mean ice crystal diameter
$\overline{s}_0$	ice crystal diameter immediately after freezing
s_{max}	maximum ice crystal diameter as defined in Equation 15
s_e	equivalent ice crystal diameter
t	time
T_F	freezing temperature
T_p	plate freezer temperature
T_∞	bulk tunnel coolant temperature
U	velocity of dendritic ice crystal growth
V_i	total ice volume in a frozen product
X	distance of freezing front from surface of product
α_1	thermal diffusivity of frozen phase
λ	parameter defined by Equation 9
ρ_1	density of the frozen phase
σ	standard deviation of ice crystal sizes

References

Arbuckle WS (1986) Ice Cream, 4th edition. AVI Pub Co Westport, Connecticut USA

Bald WB (1973) Cryogenic heat transfer at Oxford, Part 1, Nucleate pool boiling. Cryogenics 13, 8, 457-469

Bald WB (1986) On crystal size and cooling rate. J Microsc 143, 1, 89-102

Bald WB (1987) Quantitative Cryofixation. Adam Hilger, Bristol

Bald WB (1988) Theory of rapid freezing. Inst Phys Conf Ser No 93 vol 3, Ch 1 Eurem 88, York

Bevilacqua A, Zaritzky NE and Calveco A (1979) Histological measurements of ice in frozen beef. J Food Technol 14, 237

Calvelo A (1981) Recent studies on meat freezing. In: Lawrie R (ed) "Developments in Meat Science" vol 2 Applied Science Publishers, London, New Jersey

Diller KR (1990a) A note on opitimal techniques of rapid cooling for low temperature preparation of electron microscopy specimens. Cryo-Letters to be published

Diller KR (1990b) Coefficents for solution of the analytical freezing equation in the range of states for rapid solidification of aqueous systems. Trans ASME, J Heat Transfer to be published

Frank FC (1950) Radially symmetric phase growth controlled by diffusion. Proc Roy Soc A 201 pp 586-599

Gaertner RF (1963) Distribution of active sites in the nucleate boiling of liquids. Chem Eng Prog Symp Series 59, pp 52-61

Jones H (1986) Science and technology of the undercooled melt. NATO ASI series Lancaster Martinus Nijhoff 156-185

Luyet B and Rapatz G (1958) Patterns of ice formation in some aqueous solutions. Biodynamica 8, pp 1-68

Martino MN and Zaritzky NE (1988) Ice crystal size modifications during frozen beef storage. J Food Science 53, 6, 1631-1638

Meryman HT (1966) Cryobiology, Ch 1 Academic Press, London

Riehle U and Hoechli M (1973) The theory and technique of high pressure freezing. EL Benedetti and P Favard (eds) Freeze-etching techniques and applications. Paris Societé Francaise de Microscopie Electronique

Smith CE and Schwartzberg HG (1985) Ice crystal size changes during ripening in freeze concentration. Biotech Prog pp 1, 2, 111-120

Chapter 6

The Effect of Polymers on Ice Crystal Growth

C. Holt

Introduction

Ice cream and other frozen desserts are frequently subjected to elevated temperatures during distribution. At elevated temperatures populations of ice crystals increase in average size due to recrystallisation. This detracts from the quality of the ice cream since ice creams which contain large ice crystals are perceived as coarse-textured and icy.

Small quantities of hydrocolloid stabilisers are normally added to ice creams. Stabilisers improve the body and smoothness of ice cream and it is sometimes claimed that they can reduce the rate of recrystallisation of ice on storage. This paper is a review of the published literature on the effect of stabilisers on the growth of ice crystals.

Effect of Polymers on Ice Growth Velocity

Shipe *et al.* (1963) studied how the freezing behaviour of milk and sucrose solutions was affected by the addition of gelatin, sodium alginate, agar, guar gum and carrageenan. The solutions were supercooled by 2.5°C in a Fiske cryoscope. The solutions were then seeded and the temperature was recorded at 15 s intervals for at least 5 min. On seeding the solution, the temperature began to rise as the latent heat was released and the temperature then remained constant as the ice continued to form. The temperature then fell again when the rate of heat removal exceeded the rate of latent heat release.

The addition of stabilisers to the solutions reduced the rate of temperature rise after seeding and prolonged the freezing plateau. The effect became more pronounced as the concentration of the stabilisers was increased. Although, generally, increasing the viscosity prolonged the freezing curve, viscosity was not directly controlling the freezing since different stabilisers gave the same freezing behaviour at different viscosities.

Rapatz and Luyet (1972) studied the growth rate of single ice crystals in various concentrations of gelatin. Microscope slides were prepared by heating the gelatin solutions, placing a drop on a slide and covering it with a cover-slip. Spacers 0.1 mm thick were used to control the specimen depth. The slides were immersed in well-stirred cooling baths and the growth of single ice crystals was observed. Increasing the gelatin concentration over the range 1% to 40% produced a very large reduction in crystal growth rate (Table 6.1). Growth rates were complicated by the fact that different shapes grew under different conditions but within each shape there was a large effect of gelatin concentration.

Blond (1985) measured the freezing point of solutions of carboxymethyl cellulose (CMC), low methoxy pectin, polyvinyl alcohol and gelatin. The freezing point depression was more than that predicted by Raoult's law although it was still small. For example, a 10% gelatin solution had a freezing point of -0.2°C. The evolution of the gel structure in gelatin which accompanies ageing resulted in further depression of the freezing point.

Table 6.1. Growth rates of ice crystals in gelatin solutions (after Raptz and Luyet 1972)

Type of ice formation	Conc %	Temp °C	Growth rate mm/min	Conc %	Temp °C	Growth rate mm/min
I Rosettes or Lobes	30	- 4°	0.0046	40	-10°	0.0044
	30	-10	0.0078	40	-15	0.0068
	30	-15	0.0138	40	-30	0.046
II Evanescent Spherulites	30	-30	24	40	-60	.53
	30	-40	36	40	-70	.96
	30	-50	52			
	30	-60	78			
	30	-70	103			
III Irregular Dendrites	1	-10	750	1	- 4	194
	3	-10	264	3	- 4	30
	10	-10	4.8	10	- 4	0.25

This author went on to measure the linear crystallisation velocity in polymer solutions and gels (Blond 1988). The measurements were carried out in a U-tube immersed in a cooling bath several degrees below the freezing temperature of the solution. Freezing was initiated by introducing an ice crystal at one end of the U-tube. Two thermocouples were fitted in the tube, one near each end. When the ice front passed the junctions a rise in temperature was observed and the velocity of the crystallisation was calculated from the time interval.

The propagation rate of the ice front was greatly reduced when the concentration of CMC was increased up to 5%. Solutions of high and low viscosity CMC in varying proportions were used to produce a wide range of viscosities. However the velocity of the ice front was reduced only slightly when the viscosity increased by a factor of 500 which suggests that viscosity is not the controlling factor.

Gelled systems were much more effective in reducing the rate of crystallisation than ungelled solutions. When low methoxy pectin was gelled by the addition of calcium ions the propagation rate was reduced from 11.35 mm s^{-1} to 7.23 mm s^{-1}. The propagation rate was lower in gelatin gels when the gels had been matured for a longer time at higher temperatures i.e. when the gel contained a larger number of junction zones.

Blond (1988) points out that viscosity is a macroscopic property and the water is not truly immobilised but can migrate freely through a gel structure. The slowing down of the crystal front is a mechanical effect which is imposed by the polymer molecules. The effectiveness of the hindrance depends on the concentration and flexibility of the molecules and the rigidity of the gel matrix.

Budiaman and Fennema (1987a) measured the linear crystallisation velocity in solutions of a number of hydrocolloids including carrageenan, gelatin, CMC, guar gum, LBG and sodium alginate. The solutions were placed in U-tubes which were immersed in a bath at temperatures ranging from -3°C to -5°C and the linear crystallisation velocity was measured over a straight section 10 cm long. The crystallisation velocity decreased as the hydrocolloid concentration increased. Guar gum and locust bean gum were the most effective in reducing the crystallisation velocity and carrageenan and sodium alginate were the least effective. The authors suggest that this may be due to the fact that guar and LBG have numerous galactose side groups whereas carrageenan and alginate lack side groups of any substantial size.

It was not possible to measure crystallisation velocities in gelatin gels. An unusual pattern of crystallisation was seen in which three rather indistinct fronts moved through the gel in succession. No such problems were encountered with carrageenan which also forms gels although at much lower cross-link density.

Sucrose was shown to be much more effective in reducing crystallisation velocity although in these experiments sucrose was added at a much higher concentration (4%) than the hydrocolloids.

The viscosities of these hydrocolloid solutions were measured with a Brookfield viscometer (Budiaman and Fennema 1987b). For any given hydrocolloid suspension the crystallisation velocity decreased as the viscosity increased. However the crystallisation velocity differed among hydrocolloid solutions adjusted to the same viscosity indicating that viscosity is not the important factor.

On the basis of these studies Budiaman and Fennema state that the beneficial effects of hydrocolloids on the texture of frozen desserts may arise from some attribute other than control of crystal size.

Muhr and Blanshard (1986a) studied the effect of polysaccharide stabilisers on the growth velocity of ice crystals in tubes and spherical flasks. The flask method was used to avoid the effect of the tube wall on the growth behaviour in some gels - an effect which was also seen by Budiaman and Fennema (1987a). Stabilisers in solution reduced the growth rate of ice in sucrose solutions by a factor of between one third and two thirds. They did not greatly affect the morphology of the ice which formed. The effectiveness of the stabilisers was in the order: Xanthan gum, guar gum, locust bean gum, alginate and courlose.

Stabilisers which formed gels had a much greater effect on ice growth rates and ice crystal morphology even when the rigidity of the gel was not very great. Gelatin was the most effective despite the fact that it gave a softer gel than the other stabilisers (agar and alginates). Increasing the gel strength further reduced the ice growth rate. It was concluded that if a stabiliser does not diffuse sufficiently rapidly away from the ice -water interface then it will exert an influence on the ice growth rate and morphology. This slowing may be due to the network causing curvature of the growing ice face and reducing the freezing point.

The authors suggested a different mechanism in the case of non-gelling stabilisers. It is possible that the polymer solutions are more effective in depressing the freezing point when in the presence of sucrose. Alternatively the stabilisers might diffuse much more slowly when they are in a sucrose solution and so block growth sites on the ice crystals.

Effect of Polymers on Ice Recrystallisation

Harper and Shoemaker (1983) studied ice recrystallisation in solutions containing sucrose, corn syrups, and locust bean gum. Droplets of the solutions were placed between two coverslips and frozen by contact with solid carbon dioxide. The samples were transferred to a microscope cold stage and the temperature was cycled between -23°C and -9°C four times. Some slides were also held at -23°C for 64 min.

There was a linear relationship between the ice crystal size and the number of temperature cycles or the period of time that each sample was held at a constant -23°C. The addition of locust bean gum at levels up to 0.5% did not affect ice crystal growth (this was only tested in the 35% corn solids solution). However the type of sweetener affected the growth rate. Effectiveness in inhibition was in the order sucrose > corn syrup > fructose.

Recrystallisation of ice in stirred sucrose solutions was studied by Smith and Schwartzberg (1985). Ice crystals of diameter 0.09 mm were produced by spraying water into liquid nitrogen and sieving the resulting frozen droplets. The crystals were added to sucrose solutions and stirred for up to 5 h. High sucrose concentrations were seen to slow down ripening by increasing the viscosity and so reducing the mass transfer coefficients that control crystal growth rate in conditions of mild agitation. Relatively low concentrations of gelatin (up to 0.05%) greatly reduced the rate of ice recrystallisation by similarly reducing the mass transfer coefficient.

Buyong and Fennema (1988) manufactured butterfat ice creams with and without 250 Bloom gelatine at a level of 0.28%. The unaerated mixes were spread in 1 mm thick layers onto microscope slides and rapidly cooled. The slides were stored at -15°C for two weeks to encourage the ice crystals to grow. Specimens were prepared for electron microscopy by freeze-drying and coating with gold/palladium. The gelatin had no effect on the intial size and shape of the ice crystals and had no effect on the rate of ice crystal growth during the two-week storage period.

Effect of Polymers on Ice in Products

Cottrell *et al.* (1979) studied the effects of various polysaccharide stabilisers on the physical properties of ice cream. The ice creams were stored for 16 weeks at -35°C and then tested for appearance, taste and texture. It appeared that the seaweed extracts (agar, carrageenan and alginate) gave poor control of ice crystal growth since the ice creams containing these stabilisers became icy during the 16 weeks' storage. There was no correlation between inhibition of ice crystal growth (as measured by iciness scores) and either the water binding capacity or the viscosity of the ice cream mix.

Champion *et al.* (1982) prepared a range of ice creams containing various levels of microcrystalline cellulose. The ice creams were stored at -16°C for three months and the hardness of the ice creams was measured at intervals with a penetrometer. The ice creams became harder during storage which was believed to be due to the formation of increasingly large ice crystals. As the concentration of microcrystalline cellulose in the ice cream increased the increase in hardness was retarded. The authors suggest that the growing ice crystals are prevented from coalescing (to form larger ice crystals) by the microcrystalline cellulose which acts as a physical barrier. Wittinger and Smith (1986) made a 10% fat ice cream containing 0.15% stabiliser which comprised a mixture of guar gum and locust bean gum in various ratios. The ice creams were stored in a freezer which cycled between -9.4°C and -15°C for ten weeks. Sensory analysis was carried out at intervals and included scores for iciness. Ice creams containing high levels of guar gum became icy sooner than ice creams containing high levels of locust bean gum.

Discussion

There is ample evidence that polymers slow down the growth rate of ice and alter ice crystal shape in undercooled aqueous solutions. Several authors have noted that the reduction is greater, and the change in shape more pronounced, if the polymer forms a gel. Viscosity does not seem to be the important factor - a number of studies have shown that different stabilisers can give the same freezing behaviour at widely differing viscosities. In any case viscosity is a macroscopic property and does not reflect the diffusion rate of small molecules. Measurements have shown that, even in a gel, the diffusion of small molecules is not greatly reduced (Muhr and Blanshard 1982). So the effect of polymers on ice crystal growth is unlikely to be due to a reduction in the mobility of water.

It has also been demonstrated that polymeric materials have little effect on the nucleation of ice (Muhr and Blanshard 1986b), and very little effect on the freezing point of aqueous solutions (Blond 1985). The effect of polymers on ice crystal growth appear to be due to the rheological properties of the solution or gel. If the polymers do not diffuse away from the ice surface sufficiently rapidly then they will exert an influence on the ice. The mechanical hindrance or obstruction of the growing ice face is then increased by raising the polymer concentration, or increasing the length of the polymer molecules, or by cross-linking to form a gel.

What effect polymers have on ice crystals in ice cream or frozen food is much less clear. Polymers can slow down growth rates in stirred sucrose solutions (Smith and Schwartzberg 1985) but this is hardly relevant to foods. Harpur and Shoemaker (1983) saw no effect of locust bean gum on recrystallisation in sugar solutions. Buyong and Fennema (1988) found that gelatin had no effect on the initial size or the growth rate of the ice crystals in ice cream. Some other workers have claimed to see an effect of stabilisers on ice crystal growth in products but they did not observe the crystals directly but relied on taste panel scores of iciness or on hardness measurements. So whilst it is widely believed that stabilisers interfere with ice crystal growth in ice cream and frozen products, it is difficult to find any real evidence to support this.

References

Blond G (1985) Freezing in polymer-water systems and properties of water. In: Properties of water in foods, D Simatos and JL Multon (eds) Martinus Nijhoff, 531-542

Blond G (1988) Velocity of linear crystallization of ice in macromolecular systems. Cryobiology 25, 61-66

Budiaman ER and Fennema O (1987a) Linear rate of water crystallization as influenced by temperature of hydrocolloid suspensions. J Dairy Science 70: 534-546

Budiaman ER and Fennema O (1987b) Linear rate of water crystallization as influenced by viscosity of hydrocolloid suspensions. J Dairy Science 70: 547-554

Buyong N and Fennema O (1988) Amount and size of ice crystals in frozen samples as influenced by hydrocolloids. J Dairy Science 71: 2630-2639

Champion SA, Phillips GO and Williams PA (1982) The effect of microcrystalline cellulose on the organoleptic properties of ice cream. Prog Fd Nutr Sci 6, 361-366

Cottrell JI, Pass G and Phillips GO (1979) Assessment of Polysaccharides as Ice Cream Stabilisers. J Sci Food Agric 30: 1085-1088

Harper EK and Shoemaker CF (1983) Effect of locust bean gum and selected sweetening agents on ice recrystallisation rates. J Food Science 48: 1801-1803, 1806

Muhr AH and Blanshard JMV (1982) Diffusion in gels, Polymers, 23: 1012-1026.

Muhr AH and Blanshard JMV (1986a) Effect of polysaccharide stabilizers on the rate of growth of ice. J Food Technol 21: 683-710

Muhr AH and Blanshard JMV (1986b) Effects of polysaccharide stabilisers on the nucleation of ice. J Food Technol 21: 587-604

Rapatz G and Luyet B (1972) Patterns of ice formation and rates of ice growth in gelatin solutions. Biodynamica 11: 117-123

Shipe WF, Roberts WM and Blanton LF (1963) Effect of ice cream stabilizers on the freezing characteristics of various aqueous systems. J Dairy Science 46: 169-175

Smith CE and Schwartzberg HG (1985) Ice Crystal Size Changes During Ripening in Freeze Concentration. Biotechnology Progress 1: 111-120

Wittinger SA and Smith DE (1986) Effect of sweeteners and stabilizers on selected sensory attributes and shelf life of ice cream. J Food Science 51: 1463-1470

Chapter 7

The Effect of Freezing on some Properties of Quorn Myco-Protein

G. W. Rodger and R. E. Angold

Introduction

Freezing as a method of protecting foods against microbial spoilage has long been an accepted technology in the food industry. However, as the practice became routine, it was observed more and more that conditions which prevented microbial spoilage of the food did not necessarily prevent other "events" from occurring which led to sensory quality deterioration. For example, lipid oxidation leading to off-flavour development in fatty foods (Hardy 1980), aggregation of myofibrillar protein in fish muscle effecting textural changes (freeze denaturation) (Suzuki 1981), and ice crystal size disproportionation (Ostwald ripening) causing the sensation of "iciness" in ice cream (Blanshard and Franks 1987)).

As the reported number of such observations grew, it became apparent that treating the freezing process only as the conversion of water in the food to ice was much too simplistic an approach. It is now well accepted that

(i) the "rate" at which freezing occurs (which influences ice crystals size distribution (Lawrence *et al.* 1986)),

(ii) the time and temperature of subsequent frozen storage via their influence on reaction kinetics, and

iii) the influence of temperature on the material properties of frozen products which affect their processability in that state (Sheard *et al.* 1989) (e.g.) whether frozen meat undergoes brittle or plastic fracture during beefburger manufacture) are all significant factors in determining the final sensory quality of frozen foods.

As the causal connections between these physical/chemical events and the resulting effects on processing and sensory quality of the food began to be established, methods of counteracting the deleterious effects were devised, e.g. the use of:

(a) polymeric stabilisers to reduce the rate of Ostwald ripening in ice cream (Blanshard and Franks 1987);
(b) cryo-protectants to reduce the effects of freeze-denaturation in fish (Tsuchiya *et al.* 1975); and
(c) the general realisation that the lower the temperature of storage, the less the degree of quality change.

In some instances, however, the presence of ice crystals, and their subsequent growth, have been used as a process aid. For example, the soya bean curd product, Tofu, is converted into a more textured product, Kori-Tofu, via freezing (Wolf and Cowan 1971), and in the early/mid 70's a series of patent applications were made exploiting the use of ice crystal growth in imparting meat like texture to finely comminuted protein "pastes" (General Foods (1978) British Patent Spc. 23345/77, Nishin Flour Milling Co (1973) Jap. Patent JA 7 321, 502, Unilever Ltd (1979) British Patent 1, 544, 906).

It is intended that the examples given above introduce the concept that the influences of freezing on the quality of foodstuffs can result from different effects. The area to be discussed in this paper will concentrate on how freezing can affect the physico-chemical properties of the *raw material* (myco-protein) used in the manufacture of a range of novel protein based foods currently sold by several large retail outlets in the UK.

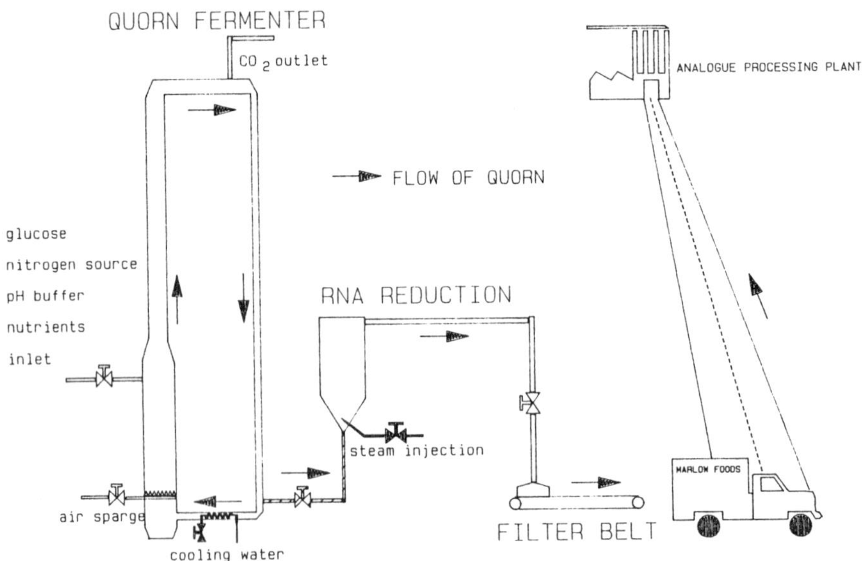

Fig. 7.1. Schematic showing production of Quorn myco-protein raw material.

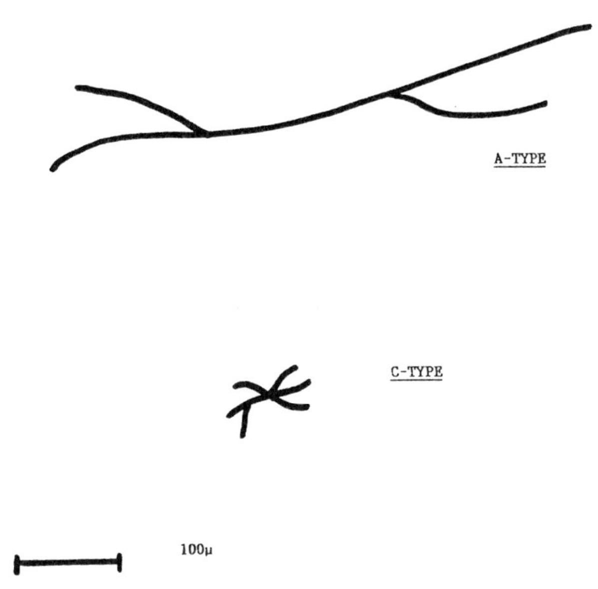

Fig. 7.2. Typical morphologies of 'A' and "C: type hyphae of *Fusarium graminearum*.

Myco-Protein as a Raw Material

"Quorn" myco-protein, as a raw material, comprises a mat of filtered fungal hyphae (RNA reduced) derived from the continuous fermentation of a strain of *Fusarium graminearum* via the process outlined in Figure 7.1. The dimensions of the hyphae are approximately 3 μm-4 μm diameter, and 300-600 μm in length (Fig. 7.2) The hyphal wall contains mainly chitin and ß glucans, and the intra hyphal space denatured cytoplasmic proteins, as a result of the RNA reduction process which will be discussed later (Fig. 7.3). On a dry basis, the analysis of myco-protein is 50% Protein, 1.1% RNA, 12.4% Lipid, 3.3% Ash, 2.7% H_2O soluble sugars , and 24.6% non-digestible fibre .

A WHO/FAO recommendation exists that the contribution of RNA to the diet from a novel food source should be limited to 2g per day. Since the "live" organism contains about 8% RNA *ex fermenter*, then not attempting to reduce this level would limit the level of ingestion of myco-protein to approximately 20g/day dry weight. Thus Anderson and Solomons (1984) adopted the heat shock system devised by Maul *et al.* (1970), which results in the treated organism having less than 1% RNA.

After the filter cake has been harvested (thickness = 3 mm - 4 mm) it is blast frozen and stored. When required for final product manufacture, the filter cake is thawed, then mixed with egg albumen (which binds the product during subsequent heat setting), colour if required, and flavours. This part of the production process is outlined in Figure 7.4.

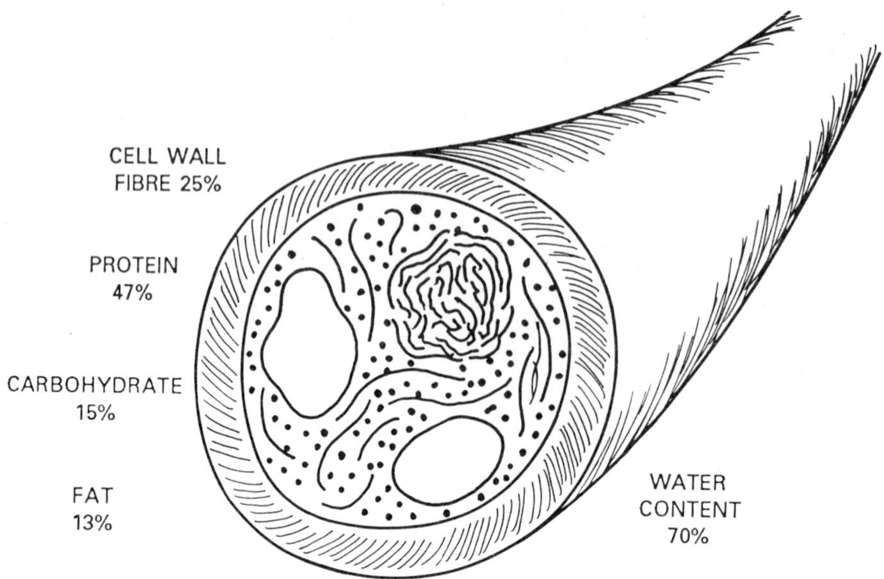

Fig. 7.3. Basic structure of myco-protein.

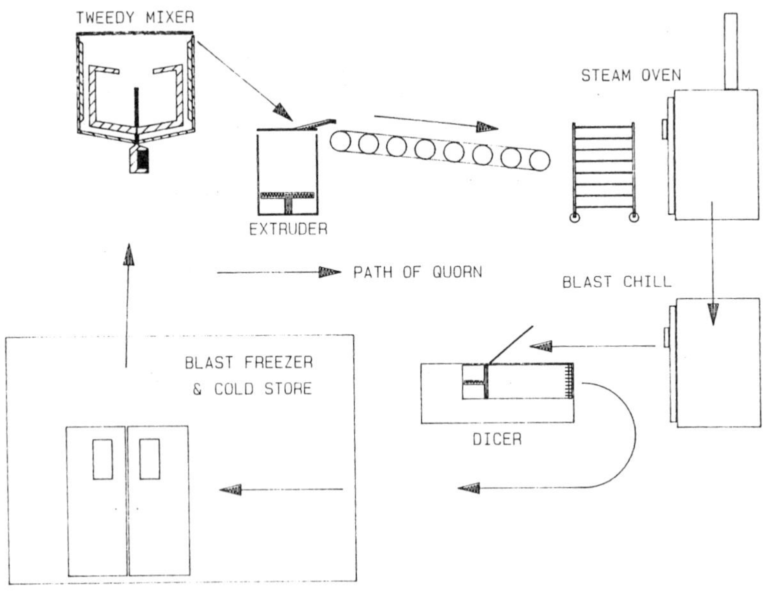

Fig. 7.4. Schematic showing final production process for Quorn myco-protein.

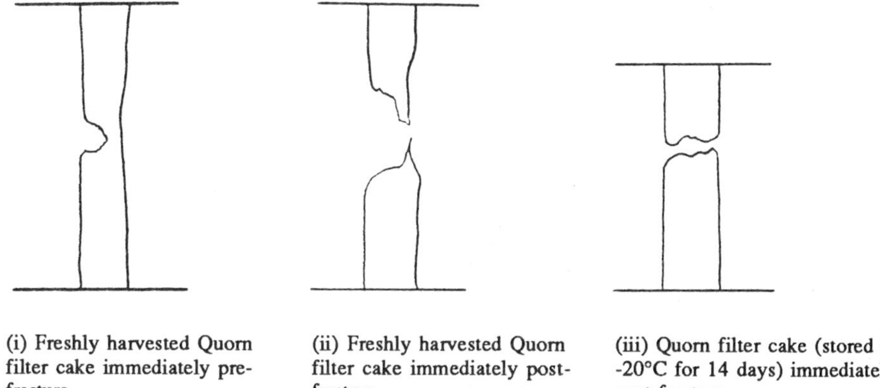

(i) Freshly harvested Quorn filter cake immediately pre-fracture

(ii) Freshly harvested Quorn filter cake immediately post-fracture

(iii) Quorn filter cake (stored at -20°C for 14 days) immediately post-fracture

Fig. 7.5. The deformation behaviour of Quorn myco-protein under tension pre and post freezing.

As stated above, however, this paper will be concerned only with the changes effected in the *filter cake* by the freezing process and subsequent frozen storage, not the final product.

Observations on the Effects of Freezing and Frozen Storage on Filter Cake Properties

Prior to freezing, the harvested filter cake has a "putty-like" consistency. If it is squeezed by hand, it flows and it is very difficult to express any fluid in this way. After normal blast freezing and storage, the behaviour of such filter cake is considerably altered. The cake now has chamois-leather type properties - if squeezed, it does not flow as before and, in this instance, water can be expressed fairly easily. In addition, the fracture properties of the sheet under tension change markedly as shown in Figure 7.5. The "fresh" material draws out and exhibits "necking", whereas the frozen, thawed sample undergoes fracture of a much more brittle nature (i.e. far less extensibility at fracture).

There are, of course, variations to this theme. In the introduction, it was stated that regarding freezing simply as the conversion of water to ice was over-simplistic, and our experience with myco-protein amply demonstrates this view as follows. The response of some properties of myco-protein to ice formation is very much freezing rate and storage temperature dependent, as indicated in Figure 7.6. If filter cake is frozen by liquid nitrogen and then stored at -20°C, then very little change in the behaviour of the system occurs. Its rheology is virtually unaffected and there is very little increase in the amount of water which can be expressed. As the freezing rate decreases and storage temperatures increase, however, the changes remarked upon above become more and more noticeable. The ultimate example of this is if filter cake is frozen very slowly (e.g. in a domestic freezer), and then stored at -5°C. In this instance, the water in the

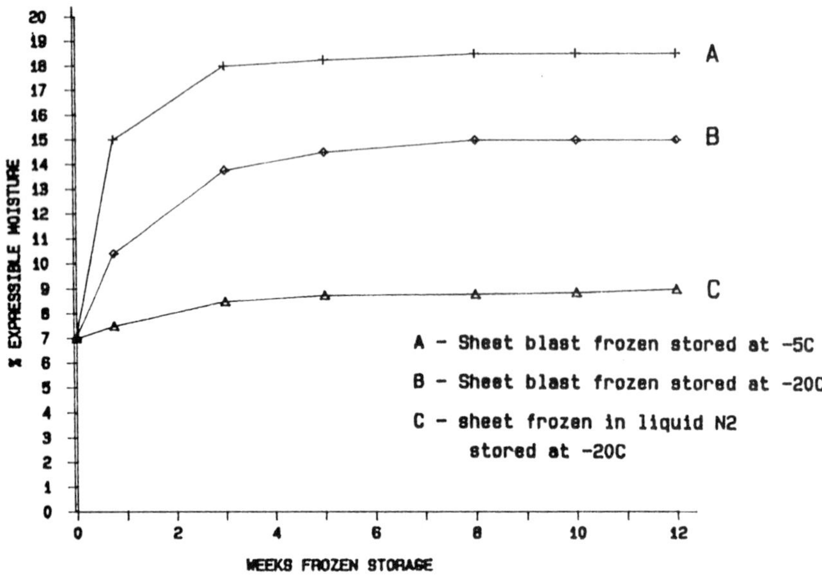

Fig. 7.6. Effect of freezing rate and storage temperature on the level of expressible moisture from Quorn myco-protein sheet.

system is held so weakly that it drips from the sheet under gravity post thawing, and the tensile strength of the cake is very much greater than "normal" frozen stored material.

Explanation of Observed Effects

Since our system comprises "heat killed" hyphae, then any effects induced by freezing are unlikely to be enzyme-mediated in the way that the water holding properties of muscle tissue can be influenced by processes such as post-mortem glycolysis. Similarly, we doubt that the response to freezing is a result of molecular aggregation of the kind which occurs during the "freeze-denaturation" of fish muscle proteins.

We strongly suspect that the response to freezing we observe can be explained simply by considering ice crystal growth in the system as follows. However, in doing so, we believe a brief description of ice behaviour is warranted.

Freezing Rate Effects

Normally, when cellular systems freeze, ice crystal nucleation occurs extracellularly since:

(i) the concentration of dissolved solutes is generally less there, which means that the freezing point will be higher; and
(ii) since cells form a compartmentalised system, the chances of nucleation occurring within a cell, leading to subsequent crystal growth throughout the bulk, are smaller than nucleation occurring in the extracellular space (Taylor 1986).

As the system cools further, water will preferentially crystallise onto existing crystals, rather than form new nuclei (Franks 1985). The net result of this is that as the extracellular water freezes, the concentration of dissolved solutes will increase until it becomes greater than the cellular concentration. At this point, osmotic pressure effects will "pull" water out of the cells, eventually to crystallise extracellularly.

In addition, the number of nucleation centres which are created (which in turn will influence the number and ultimately the size of ice crystals) is dependent on freezing rates, degree of undercooling, and the type and concentrations of dissolved solutes (Franks 1985).

Storage Effects

In most frozen systems of any bulk, a range of ice crystal sizes exists, and during storage, even at constant temperatures, the smaller ice crystals decrease in size as larger ones increase, since smaller crystals have higher surface energies than large. Since this process depends in part upon the ability of water to diffuse from the small to the large ice crystals, the kinetics will be influenced by the presence of dissolved solutes, which will increase the viscosity of the residual aqueous phase as more and more water is progressively frozen out.

Why Ice Crystals and their Size affect Filter Cake Preparation

In a filtered mass such as myco-protein the mechanical properties of this system depend on two contributions - the first is the properties of the individual elements which comprise the mass, and the second is the structure of the mass which determines the degree of bonding/entanglement between fibres (Page 1969). In the case of myco-protein, ice crystal growth can be implicated in both mechanisms as follows - if the cells are osmotically dehydrated during freezing, and on subsequent thawing do not achieve their former volume, then it can be argued that the mechanical "strength" of the cell should increase since it effectively becomes more "concentrated".

Correspondingly, given the nature of the fibrous mass, ice crystal growth will tend to concentrate the hyphae between the faces of the ice crystals created, thus increasing the degree of inter-fibre bonding, which in turn will lead to an increase in the mechanical strength of the filter cake via increased frictional effects.

Why Filter Cake Properties are Important

Since we consider our final products as composite systems, then purely from a materials science view point, we can state that the following properties of the system are important:

Dispersion state of the included phase (hyphae) i.e. is it monodisperse or do the hyphae exist as aggregated "clumps"?
Hyphal morphology (diameter, length);
Hyphal phase volume;
Hyphal flexibility/rigidity i.e. visco-elastic properties;
Matrix (i.e. binder) phase volume;
Matrix visco-elastic properties (i.e. binder gel properties).

As far as freezing and its possible effects are concerned, then immediately we can see that:

i) the mechanical properties of the filter cake will influence the degree of filter cake dispersion during mixing;
ii) if the hyphae undergo irreversible shrinkage resulting from freeze-induced osmotic effects then both their phase volume in the product and visco-elastic properties will be altered;
iii) the amount of extracellular water present will influence the concentration, hence visco-elastic properties, of the "binder".

Conclusion

At present, we are investigating the contribution of variances in these raw material characteristics to what is, after all, the most important property of all - the sensory quality of the final product. However, the results of these studies must wait for another time. The authors hope to have provided more evidence in this paper that the suitability of raw materials for a particular production process can be dependent on its response to freezing conditions used to prevent microbial spoilage.

References

Anderson C, Solomons GL (1984) Primary metabolism and biomass production from Fusarium. In: Moss MO, Smith JE (eds) The applied mycology of Fusarium, Cambridge University Press Cambridge pp 231-250

Blanshard JMV, Franks F (1987) Ice crystallisation and its control in frozen food systems. In: Blanshard JMV, Lillford PJ (eds) Food structure and behaviour, Academic Press, London, pp 51-65

Franks F (1985) Biophysics and biochemistry at low temperatures, Cambridge University Press Cambridge

Hardy R (1980) Fish Lipids Part 2. In: Connell JJ (ed) Advances in fish science and technology, Fishing News Books, Farnham, Surrey, England pp 103-111

Lawrence R, Consolacion F, Jelen P (1986) Formation of structured protein foods by freeze texturisation. Food Technology 40: 3 77-82, 90

Maul SB, Sinskey AJ, Tannenbaum SR (1970) New process for reducing the nucleic acid content of yeast. Nature 228: 181-183

Page DH (1969) A theory for the tensile strength of paper. Tappi 52: 674-681

Pomeranz Y (1985) Some traditional foods. In: Functional properties by food components, Academic Press, London pp 443-464

Sheard PR, Jolley PD, Hall LD, Newnan PB (1989) Technical Note: The effect of temperature and raw material on the size distribution of meat particles pre-broken by grinding. Intl J Fd Sci and Tech 24: 421-427

Suzuki T (1981) Changes in fish protein during storage and processing. In: Suziki T (ed) Fish and krill protein - process technology. Applied Science Publishers Ltd, London pp 31-56

Taylor MJ (1986) Physico-chemical principles in low temperature biology. In: Grant BWW, Morris GJ (eds) The effects of low temperatures on biological systems, Edward Arnold, London pp 3-72

Tsuchiya Y, Nonomura Y, Matsumoto JJ (1975) Prevention of freeze denaturation of carp actomyosin by sodium glutamate. J Biochem 77: 853-862

Wolf WJ, Cowan JC (1971) Soybeans as a food source. Buttworths, London

Chapter 8

Microscopical Methods for Examining Frozen Foods

A. J. Wilson

Introduction

Most of the foods we eat are structurally heterogeneous. Naturally-derived animal and vegetable foodstuffs are composed of a complex assortment of cells and tissues. Likewise, processed foods - which could be mistakenly considered as relatively uniform, homogeneous materials are in fact heterogeneous mixtures, usually in the form of suspensions, emulsions or foams.

Our concern with the study of frozen food structure is well justified since microstructure, coupled with flavour, are undoubtedly the major characteristics of foods which render them either enjoyable or unpleasant to the consumer palate.

When highly structured materials such as foodstuffs are subsequently frozen and then stored at low temperature their composition takes on additional degrees of complexity due to freezing of part of the liquid water and oily components and the formation of new crystalline masses.

Their structural complexity is increased by the initial freezing method and the subsequent storage time/temperature regime, both of which are reflected in the size and shape of ice crystals in the foodstuff.

The effect which accompanies this important phase transition can be usefully employed as part of the manufacturing process for certain foodstuffs, to impart new mouth-textural characteristics (food texturisation) (see Rodger and Angold Chapter 7 this volume).

If a foodstuff is frozen, some of the water separates out as ice crystals while the remaining solutes and organic molecules become concentrated in the space between them, commonly referred to as the eutectic space. Meryman (1956) in a paper on the subject, stresses..... "the single most important and fundamental concept in biological freezing is that, regardless of the mysterious complexity of the biological matrix, freezing represents nothing more than the removal of pure water from solution and its isolation into biologically inert foreign bodies, the ice crystals. All the biochemical, anatomical and physiological sequelae of freezing are directly or indirectly the consequence of this single physical event."

During the relatively slow freezing process involved in preparing frozen foods, rather large (1 μm- 100 μm) ice crystals tend to form and the proteinaceous component of the food will be subjected to the concentrated eutectic mixture of its solvents and salts for the duration of the storage period. Consequently, there is a high probability that further structural change will occur in the form of aggregation of protein molecules resulting in denaturation.

A greater understanding of what happens to foodstuffs which are first frozen then stored and subsequently thawed, can only be achieved by careful investigation of their structure, particularly by the use of light and electron microscopes.

This article reviews the microscopical methods both available and amenable to the investigation of frozen food structure and highlights those new methods which could be potentially valuable.

Why Examine Frozen Food Structure?

There are a number of reasons why structural examinations of frozen food are both valuable and important:

i) to monitor ice crystal size;
ii) to monitor ice crystal shape;
iii) to monitor the extent of microstructural damage to the cells and tissues resulting from the freeze/store/thaw cycle;
iv) to monitor any likely re-distribution of solutes;
v) to establish the degree of heterogeneity of the foodstuff and draw important correlations between microstructure and mouth texture;
vi) to monitor the freezing process directly while it is happening (cryo-stage light microscopy see Chapter 14 this volume) and establish the exact temperature at which ice nucleation occurs.

Methods of Observation Available

Foodstuffs were some of the first specimens to be selected for observation by the early microscopists of the 17th and 18th centuries and subsequently microscopy has proven to be a technique of unsurpassed value in food technology. It is only in the last thirty or so years, however, that the full potential of electron microscopy (EM) has been recognised (Vaughan 1979, Holcomb and Kalab 1981). More recently, it is only as a result of the development of ancillary methods such as cryo-preparation (Robards and Sleytr 1985) and high pressure SEM (Shah 1990) that the reliability of EM observations have been improved.

Surprisingly little is documented on the application of microscopy - particularly electron microscopy - to the structural investigation of frozen or once-frozen foods despite several methods being amenable. These methods include:

i) direct macro-photography
ii) light microscopy
iii) transmission electron microscopy
iv) scanning electron microscopy
v) other potentially-useful methods

The clear advantages of macrophotography with its magnifications up to about 10X are the simplicity and speed of recording coupled with the relatively low cost of the equipment needed. The disadvantages, however, lie in the restricted useable magnification range, the small depth of field, the limited resolving power of the instrument and the low inherent contrast in most untreated frozen foods. Also, unless the specimens are freeze-dried or special low temperature environmentally-controlled cabinets are constructed, there will be considerable containment and handling problems due to the specimen surface frosting and/or subsequently melting.

The first light microscope to be equipped with a freezing stage was developed in 1897 by Molisch. The equipment consisted of an insulated box containing a freezing mixture (ice and salt). A modified microscope fitted inside the box and facilitated observations of frozen specimens. Molisch used the equipment to examine the dynamic events of ice crystal growth and also to monitor cell death in slowly frozen plant tissue.

Some of the earliest light microscope histological studies of frozen tissue were performed by Richardson and Scherubel (1908) who stabilised frozen tissue in fixatives which had been cooled to subzero temperatures where they were maintained in the liquid state by the presence of salts lowering the freezing point.

The last few years have seen significant advances in improvements to the design of cryo-stages for light microscopy with particular emphasis on controlled cooling-rates and specimen temperature control (McGrath 1987, McLennan 1989). Most are liquid nitrogen cooled and have the potential to cool to 77K. Cryo-light microscopy is now being used to investigate optimum freezing and storage conditions for frozen food and is of particular value in examining ice crystal size and rate of growth in ice cream, frozen fruit, vegetables and also meats such as fish (see McLellan *et al.* Chapter 14 this volume).

The principal problem with light microscopy of thick cryo-sections is that out-of-focus structures above and below the focal plane cause flare, which considerably reduces amplitude contrast. The scanning confocal microscope (Wilson and Sheppard 1984) overcomes this frustrating problem since the optical arrangement is such that structural detail is only resolved within a very narrow depth of focus. Consequently, the technique has the potential to derive optical section images and facilitates the non-destructive examination of three-dimensional structures in frozen food.

Transmission electron microscopy (TEM) has only recently been adopted as a useful tool for the routine examination of frozen food. The principal problem with conventional non-cryopreparation methods is that the lipids and proteins within the specimen have first to be fixed, prior to the water being removed and the tissue being infiltrated with resin. These three processes are a fundamental source of artefacts in the final image of the specimen.

Cryo-microscopical Analysis

Cryo-preparation of unfixed, untreated specimens has the advantage of circumventing the need for fixation and conventional dehydration processes since the proteins, lipids and water in the specimen are stabilised by holding them at low temperature (typically colder than 133K). In the case of a frozen foodstuff the specimen under investigation is simply rapidly cooled from the storage temperature to one which is amenable to preparation or safe observation.

A number of EM cryo-preparation methods are available for the investigation of freezing and frozen foodstuffs:

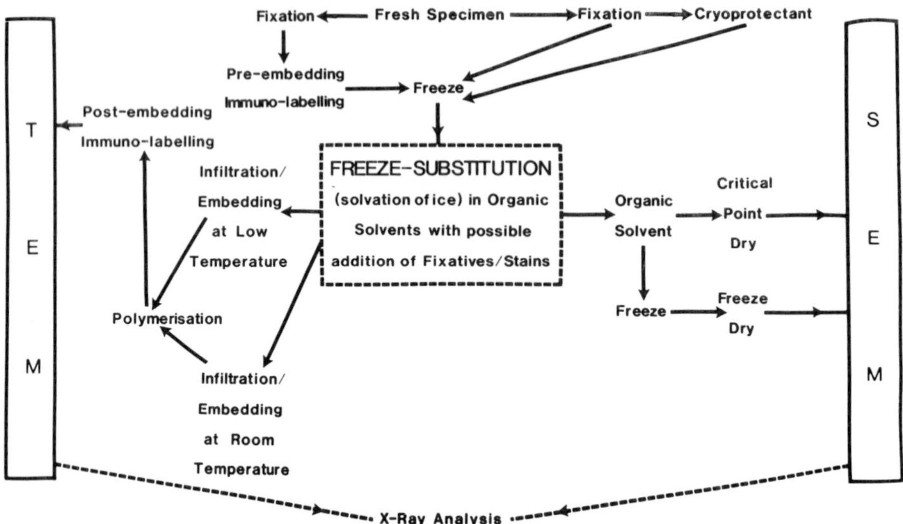

Fig. 8.1. Scheme for the cryo-preparation of specimens using freeze-substitution.

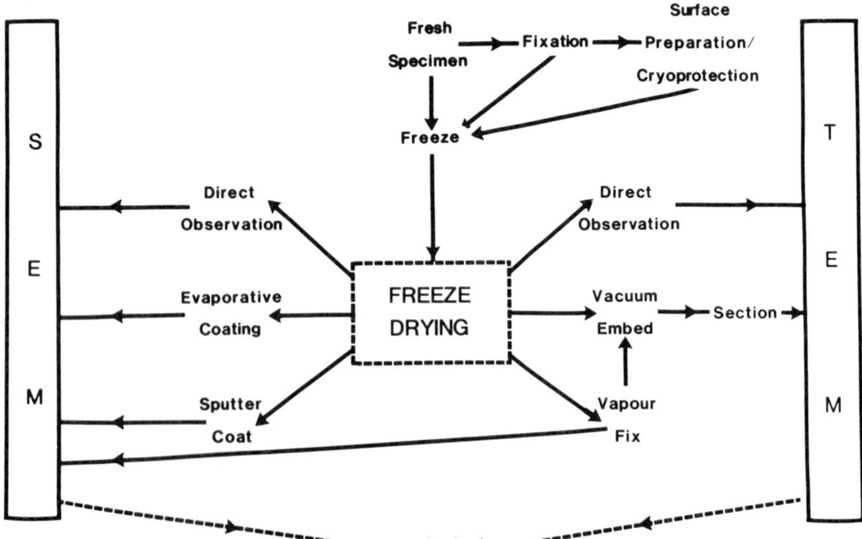

Fig. 8.2. Scheme for the cryo-preparation of specimens using freeze-drying.

i) Freeze-Substitution and Freeze-Drying

Both of these techniques have similarities in that they affect the stabilisation of frozen specimens through controlled dehydration (Figs 8.1 and 8.2).

In the case of freeze-substitution, the ice which is formed in the specimen after freezing is slowly etched away and progressively replaced by a polar solvent at low temperature. The temperature selected for the substitution to take place is always below the freezing point of the liquid contained within the specimen at room temperature (RT) (which in the case of frozen food is predominantly water) but above the freezing point of the substitution medium (methanol or acetone, Humbel and Müller 1986)

Freeze-drying involves a similar principle except ice within the specimen is sublimed under vacuum at low temperature. The temperature of the specimen and the vacuum are controlled to maintain the partial pressure of water at the specimen surface below the saturation vapour pressure of ice contained within it. Careful adjustment of the specimen temperature controls the rate of sublimation of ice from its surface and minimises ion relocalisation and structural damage. Recently, equipment has been developed which allows the freeze-drying of specimens at ultralow temperatures and high vacuum (10^{-8} to 10^{-7} mbar). This has been termed molecular distillation and facilitates the sublimation of water in its amorphous and vitreous states from rapidly cooled specimens held at temperatures below 153K (Linner and Livesey 1989).

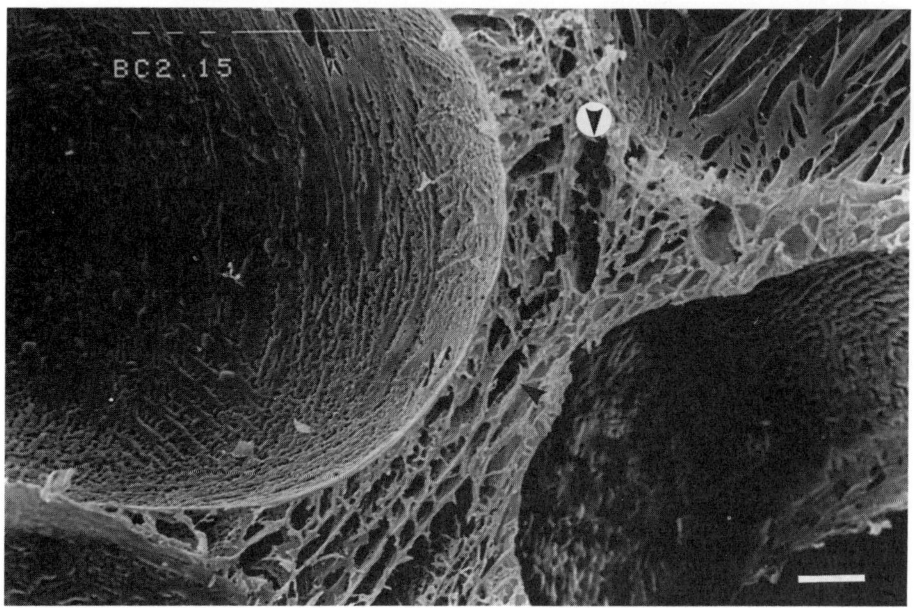

Fig. 8.3. Structure of a freeze-dried frozen foam. In the Plateau border region between the bubbles, the elongated voids which once held ice crystals are obvious (arrows).Bar = 50 µm.

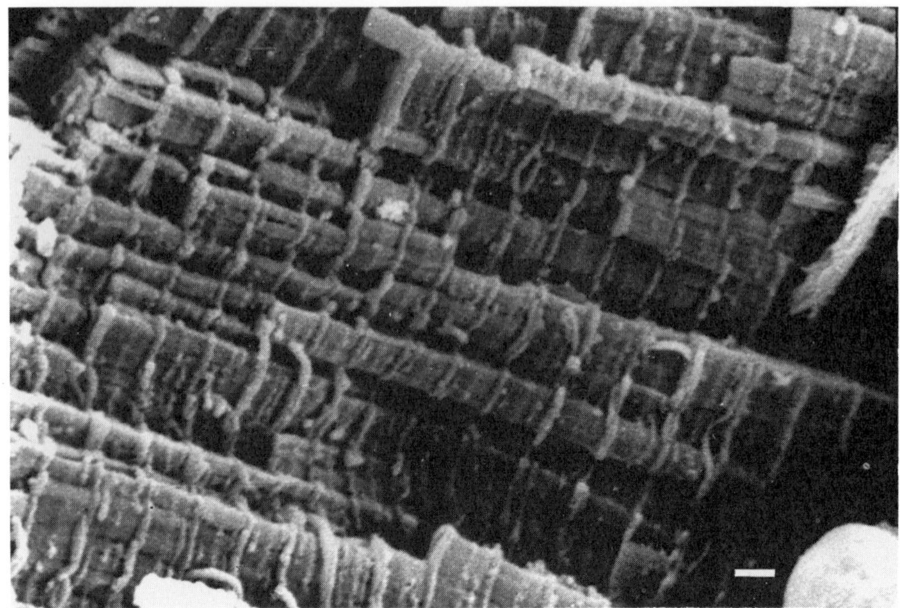

Fig. 8.4. Freeze-dried muscle fibres which have then been torn apart to reveal internal structure. The fracture generally follows the plane of the sarcolemma Bar = 1 μm.

Following either freeze-drying or substitution the dehydrated foodstuffs can be infiltrated with resin (ideally at low temperature) and polymerised prior to sectioning and staining for light microscopy (Martino and Zaritzky 1988) and TEM. Alternatively, dry specimens resulting from both techniques may be coated and then observed in the scanning electron microscopy (SEM), in the case of freeze-substituted specimens, after critical-point drying (Cohen 1973). If, after freeze-drying, specimens are warmed to room temperature and exposed to the atmosphere prior to sputter coating, then partial rehydration followed by redrying is likely to occur (Fredriksson, Bell and Lindroth 1990). Ideally, the vacuum should not be broken and the specimens coated *in situ* in the freeze-drying chamber.

Figure 8.3 demonstrates the structure of a freeze-dried beer foam. In the Plateau border region between the bubbles, the elongated voids which once held ice crystals are obvious (arrows). It is clear that the slow freezing process has resulted in considerable disruption of the Plateau border region and lamellae of the foam. Clearly the rate of freezing of the foam, its storage temperature and the regime used to thaw it will directly effect its overall stability (Lillford 1989).

In Figure 8.4 muscle has been freeze-dried, then torn apart to reveal internal structure. The fracture generally follows the plane of the sarcolemma.

Figure 8.5 illustrates the appearance of a freeze-substituted plant cell, pre-frozen in liquid nitrogen which had then been critical-point dried for SEM. The sites of ice crystals are obvious as tiny voids, those in the cytoplasm being generally larger than those in the centrally placed nucleus (single arrow) and nucleolus (double arrows).

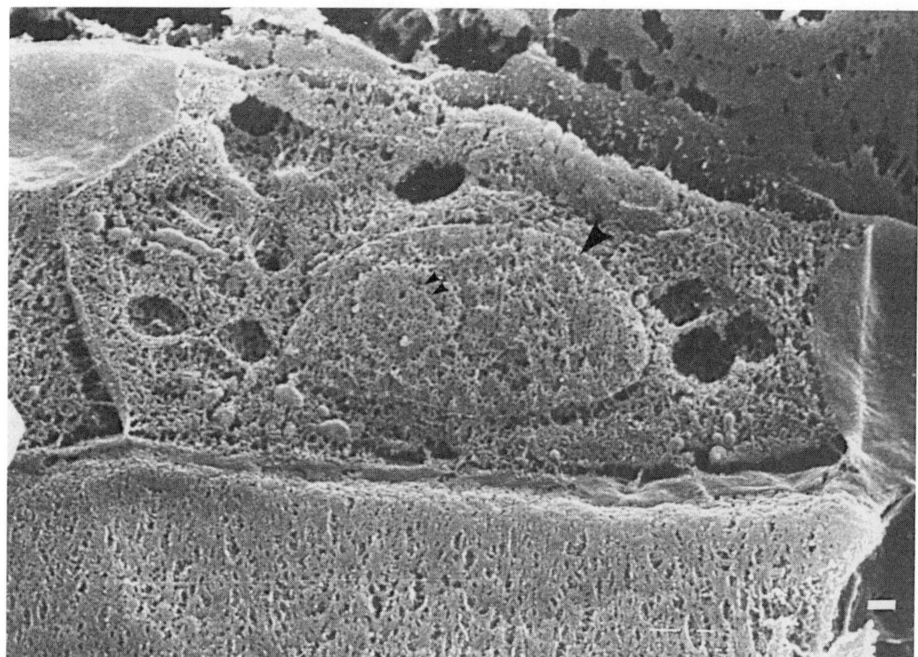

Fig. 8.5. Freeze-substituted plant cell, pre-frozen in liquid nitrogen which had then been critical-point dried for SEM. The sites of ice crystals are obvious as tiny voids, those in the cytoplasm being generally larger than voids in the centrally placed nucleus (single arrow) and its nucleolus (double arrows). Bar = 1 μm.

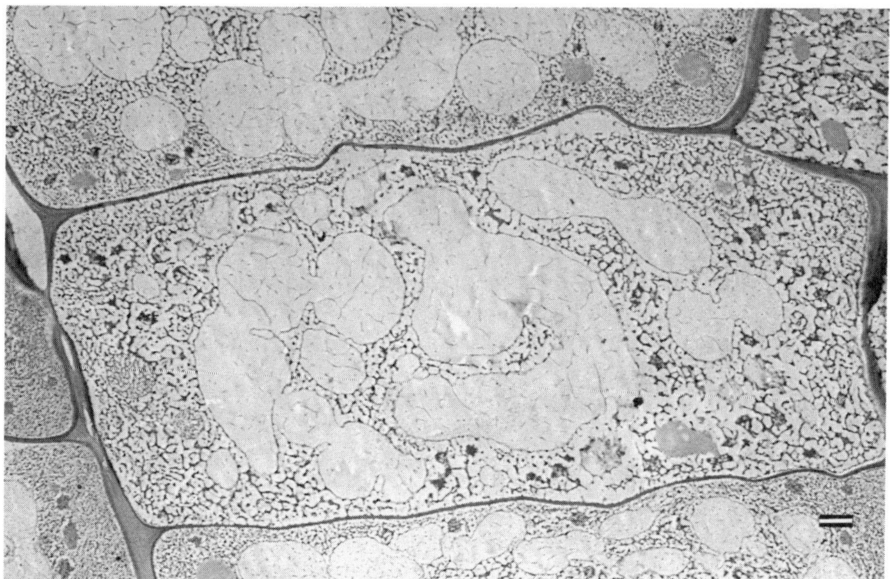

Fig. 8.6. Freeze-substituted plant cell similar to that illustrated in Figure 8.4 which was subsequently infiltrated and embedded in resin and ultrathin sectioned for TEM. Bar = 1 μm.

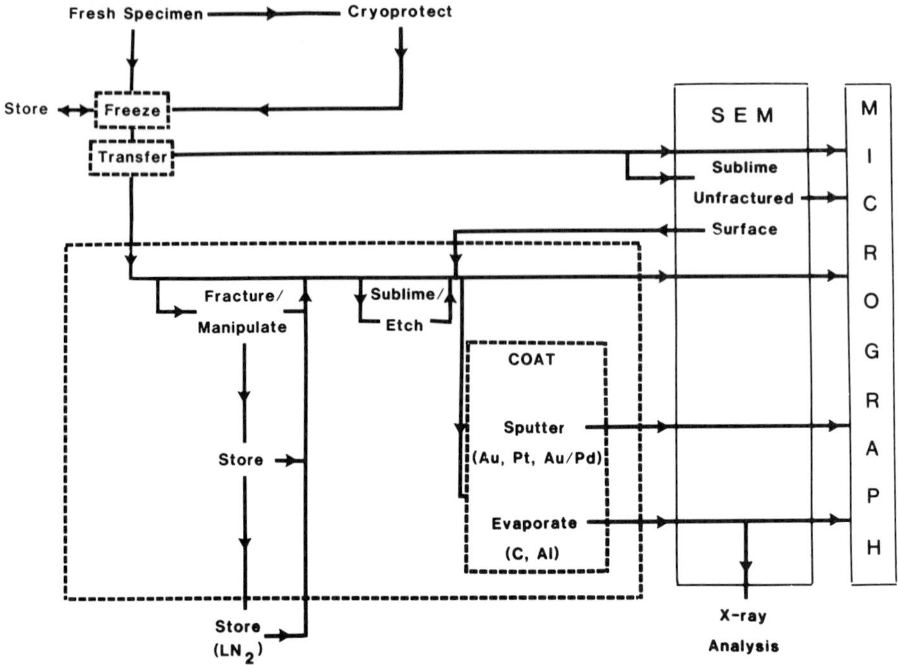

Fig. 8.7. Scheme for the cryopreparation of specimens using LTSEM.

An alternative to critical-point drying after freeze-substitution is to infiltrate the specimen with resin, polymerise and cut ultrathin sections for transmission electron microscopy. Figure 8.6 illustrates the TEM image of a similar cell to that in Figure 8.5 which was infiltrated with resin then sectioned for TEM.

ii) Low Temperature SEM (LTSEM)

This technique facilitates the observation in the SEM of pre-frozen food in the frozen hydrated condition. In addition to structural analysis, there is the option for X-ray micro-analysis of inorganic ions within the specimen.

LTSEM involves attaching a relatively large piece of frozen food (up to 10mm) onto a copper specimen stub, transferring the stub to a cold stage (typically 113K) in the evacuated chamber of the cryo-preparation unit. Inside this 'clean' environment, the specimen is fractured open to reveal a fresh surface for observation; at this point there is the option to expose sub-surface information by etching the specimen surface. This is achieved by warming the fractured specimen to a temperature of about 197K for 30 s to one minute, then immediately cooling it back to 113K (Fig 8.7).

The frozen specimen is finally coated, as is normal with SEM specimens, by sputtering with gold at low power from a thin foil target above the specimen. The frozen fractured (etched) coated specimen is then transferred, still under vacuum, to a cold stage at 113K in the scanning electron microscope where observation takes place.

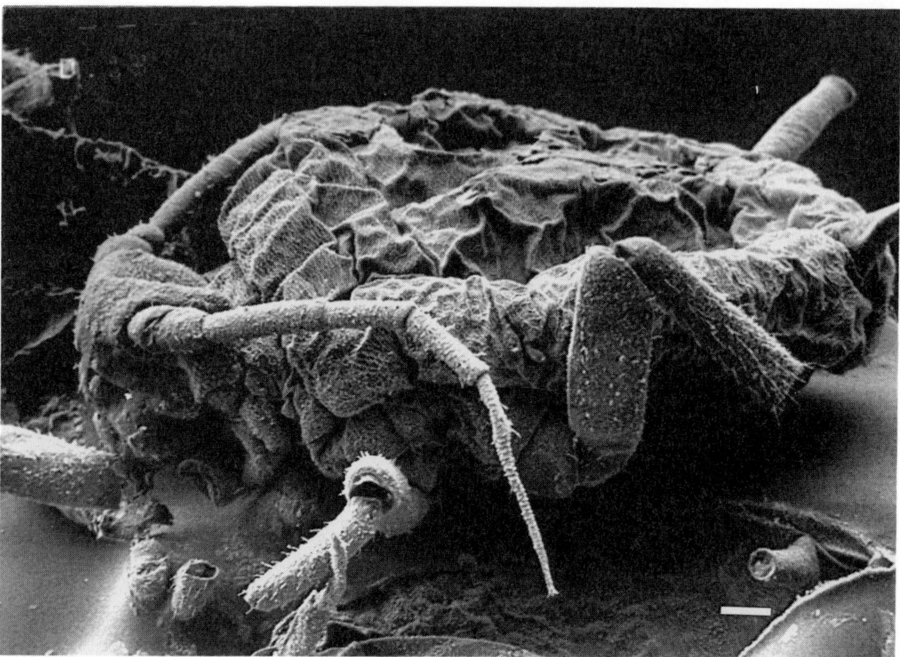

Fig. 8.8. Bean aphid which has been fixed, dehydrated and critical point dried. Note extensive collapse and shrinkage. Bar = 100 μm.

Fig. 8.9. Bean aphid prepared using LTSEM. Compare with Figure 8.8.Bar = 100 μm.

Fig. 8.10. Dendritic ice crystals. Bar = 10 µm.

Fig. 8.11. Hexagonal ice crystals. Bar = 1 µm.

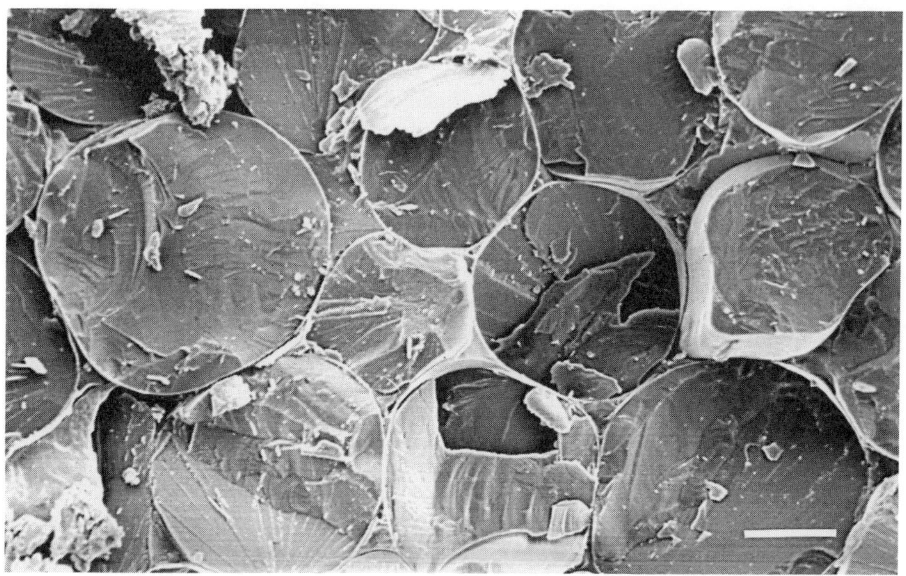

Fig. 8.12. Frozen hydrated parenchyma cells from the cortex of a plant root. Bar = 10 μm.

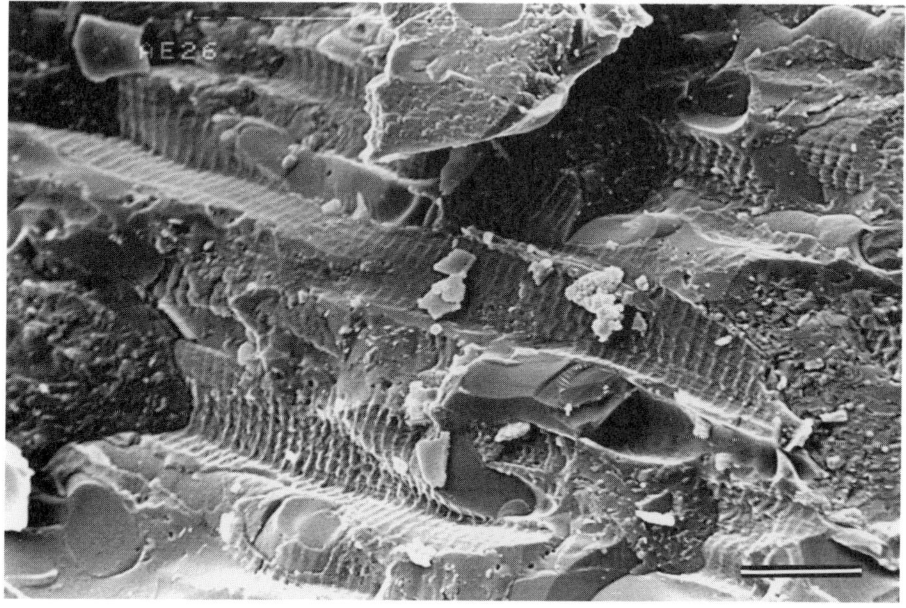

Fig. 8.13. Longitudinal cryo-fracture along the myofibril in meat. The fracture has followed the plane of the sarcolemma. Bar = 10 μm.

The immediate benefits of this technique can be illustrated by comparing the structure of the bean aphid in Figure 8.8 which had been fixed, dehydrated and critical-point dried, with the frozen hydrated preparation in Figure 8.9. Cryo-SEM is the only reliable method of investigating the structure of ice on the surface of frozen food which has been exposed to the atmosphere (Fig 8.10 dendritic ice crystals, Fig 8.11 hexagonal ice crystals). It is ideal for examining the cells of frozen vegetables (Fig 8.12) and also the structure of meat fibres illustrated in Figure 8.13 (compare with freeze-dried preparation in Fig. 8.4).

Cryo-SEM has proven value in the investigation of the structure of processed foods such as ice cream or frozen desserts which are a complex blend of foam, emulsion and suspension (Fig 8.14).

If there is a requirement for energy dispersive X-ray microanalysis to be performed after structural observation, the specimen may be either coated by evaporation of carbon or aluminium or alternatively, by sputtering with chromium. These coatings are preferable to gold, which has an X-ray peak which interferes with physiologically important elements likely to be of interest in the frozen food specimen.

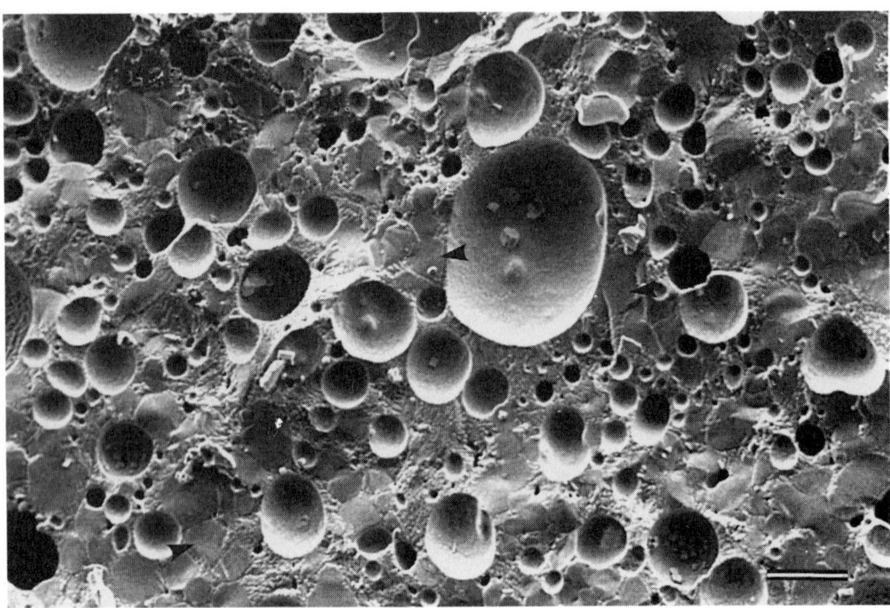

Fig. 8.14. Ice cream prepared using LTSEM. Note large ice crystals (arrows) and extensive air voids. Bar = 100 μm.

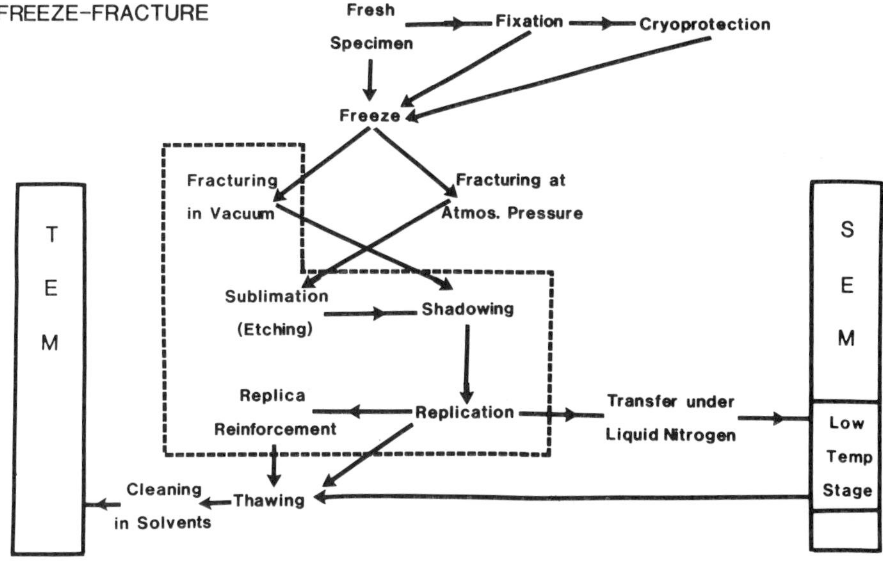

Fig. 8.15. Scheme for the cryo-preparation of specimens using freeze-fracture transmission electron microscopy.

iii) Freeze Fracture/Etch Replication

Freeze-fracture is one of the oldest techniques available to the EM cryo-microscopist, having been developed in the 1950's by Hall (1950) and subsequently Steere (1957). There are similarities between it and low temperature SEM except that freeze-fracture has the potential for high resolution observation in the transmission electron microscope. In addition, the freeze fracture replicas prepared from the frozen specimens can be stored indefinitely.

Freeze-fracture involves the replication of the surface of a freshly fractured frozen hydrated specimen (Rash and Hudson 1979, Sleytr and Robards 1977, 1982, Robards and Sleytr 1985. See Fig. 8.15).

Tiny pieces of specimen (typically 2 mm^3) are contained in metal (silver, brass or copper) holders and rapidly frozen by one of a number of means (see Steinbrecht and Zierold 1987 for review of cryofixation methods). Thc frozen specimen is then transferred to the cold stage of the freeze-fracture apparatus. At a predetermined temperature and vacuum, the specimen is fractured to prepare a clean flat surface for replication. Propagation of the fracture can be achieved by either planing the specimen with a cooled microtome blade, in which case one fracture face is formed at a desired level in the specimen. Alternatively, the frozen tissue may be snapped apart, when two (complementary) fracture faces will be formed. The disadvantage of the snap method is that the exact position of the fracture plane within the specimen cannot be easily controlled.

Fig. 8.16. An oil-in-water emulsion prepared using freeze-fracture TEM. In this sample the dispersion phase is oil (double arrows) distributed in a continuous aqueous medium and stabilised by the presence of protein (single arrows). Bar = 10 μm.

The newly-formed fracture surface is next replicated by evaporating platinum carbon from an electron beam gun or resistive heater at an angle of 45°. The resulting replica is then stabilised by depositing carbon from a source immediately perpendicular to the plane of the fracture.

After fracture and immediately prior to replication, there is the option of etching/subliming the specimen surface in order to reveal sub-surface detail.

Uni-directional shadowing does not provide contrast in all directions and indeed one large structure may obscure or overshadow smaller neighbouring ones (see Fig 8.16). One method which overcomes these problems is the technique of rotary shadowing which involves rotating the specimen during shadowing (Heuser 1981). This method is potentially very useful for examining the structure of foodstuffs which are in the form of micro-emulsions or suspensions, in conjunction with stereoscopic viewing.

The metal replica formed by this process is then separated from the underlying specimen by floating it off in a liquid which also acts as a solvent for the specimen. After a number of washes or solvent treatments to remove all traces of specimen contamination, the replica is picked up on a TEM grid prior to observation. When dry, the replicas may be stored indefinitely without deterioration.

Freeze-fracture is the only current TEM preparation method available for the high resolution investigation of liquid materials such as emulsions. In the example illustrated in Figure 8.16 the dispersion phase is oil (double arrows) distributed in a continuous aqueous medium and stabilised by the presence of protein (single arrows).

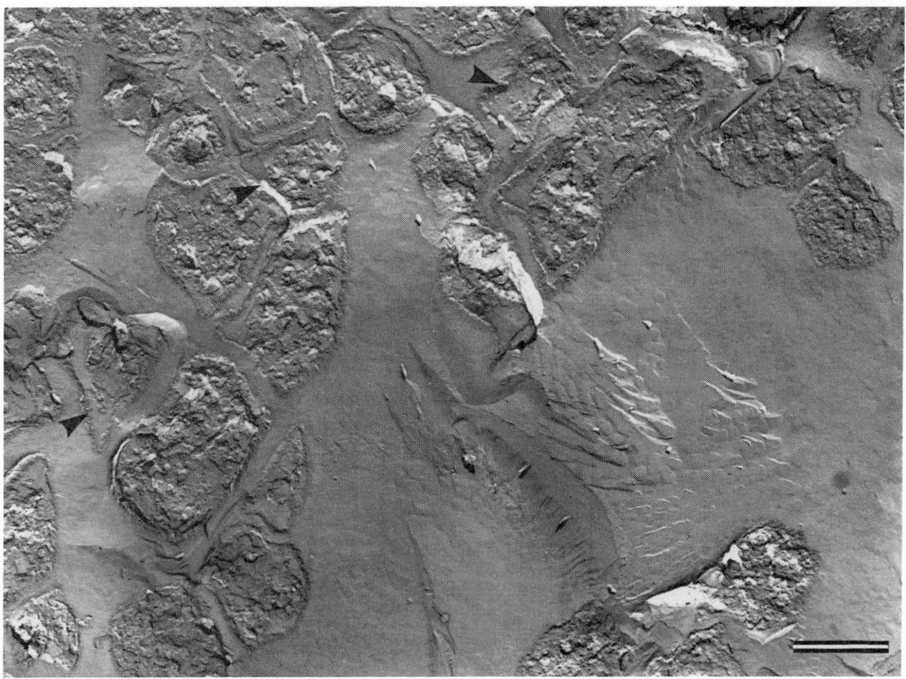

Fig. 8.17. Illustrates the appearance of a cross fracture through a bundle of frozen hyphae of the fungus *Fusarium graminearum*. As a result of freezing, the hyphae have experienced some osmotic dehydration and have lost their circular appearance (arrows). Bar = 10 µm.

Figure 8.17 illustrates the appearance of a cross fracture through a bundle of frozen hyphae of the fungus *Fusarium graminearum*. As a result of the rapid freezing method, the hyphae have experienced some osmotic dehydration and have lost their circular appearance (arrows). In addition, ice crystal growth has concentrated the hyphae by squeezing them together into bundles thereby increasing inter-fibre bonding.

Generally, foodstuffs contain both gross structures (>50 µm) and also fine detail (<10nm). Consequently freeze-fracture investigations should always be preceded by a thorough examination using SEM. The specimen for SEM observation may be either freeze-dried, or preferably, maintained in the frozen hydrated condition (LTSEM).

At the moment, two separate pieces of dedicated equipment are necessary to perform LTSEM and freeze-fracture for TEM. However, apparatus is currently being evaluated which has the potential to carry out both techniques within a single process. This promises to be a unique and exciting development not only for food technology but also for biological and materials science generally.

References

Cohen AL (1973) Scanning electron microscopy: Principles and Methods. Hayat MA (ed) Van Nostrand Reinhold Co, NY

Fredriksson B-A, Bell P, Lindroth M (1990) High performance, combined freeze-dryer and sputter coater for cryosputtering of biological samples for electron microscopy. Proc XII Int Cong EM San Francisco Press CA

Hall CE (1950) A low temperature replica method for electron microscopy. J Applied Physics 21:61

Heuser JE (1981) Preparing biological samples for stereomicroscopy by the quick-freeze, deep-etch, rotary replication technique, In: Methods in cell biology vol 22. Turner JN, (ed) Acad Press NY

Holcomb DN, Kalab M (1981) Studies of food microstructure. SEM Inc AMF O'Hare (Chicago) IL 60666-0507 USA

Humbel B, Müller M. (1986) Freeze-substitution and low temperature embedding.In: The Science of Biological Specimen Preparation for Microscopy and Microanalysis, Revel J-P Barnard T and Haggis GH (Eds) p175

Lillford PJ (1989) Structure and Properties of Solid Food Foams, In: Foams: Physics, Chemistry and Structure Wilson AJ (ed) Springer Verlag Lond

Linner JG, Livesey SA (1989) Low temperature molecular distillation drying of cryofixed biological samples. In: Low Temperature Biotechnology: Emerged Applications and Engineering Contributions. McGrath, JJ and Diller KR (eds) pp 147-157

Martino MN, Zaritzky NE (1988) Ice crystal size modifications during frozen beef storage. J Food Sci, 53, No 6, pp 1631-1637

McGrath JJ (1987) Temperature-controlled cryogenic light microscopy - an introduction to cryomicroscopy. In: Grout BWW, Morris GJ (eds). The Effects of Low Temperatures on Biological Systems. E. Arnold

McLennan M (1989) Temperature-controlled light cryomicroscopy, Lab Equip Digest (Nov)

Meryman HT (1956) Science 124 p 515

Rash JE, Hudson CS (eds) (1979) Freeze-fracture: methods, artefacts and interpretations Raven Press, NY

Richardson WD, Scherubel E (1908) J Amer chem Soc, 30 p 1515

Robards AW, Sleytr U (1985) Low Temperature Methods in Biological Electron Microscopy. Vol 10. Practical Methods in Electron Microscopy, (Ed) Glauert AM, Elsevier

Shah JS (1990) Scanning electron microscopy at high pressure. Microscopy and Analysis 17:27

Sleytr U, Robards AW (1977) Freeze-fracturing: a review of methods and results. J Microsc 111:77

Sleytr U, Robards AW (1982) Understanding the artefact problem in freeze-fracture replication: a review. J Microsc 126:101

Steere RL (1957) Electron microscopy of structural detail in frozen biological specimens. J Biophys Biochem Cytol 3:45

Steinbrecht RA, Zierold K (1987) Cryotechniques in Biological Electron Microscopy, Springer-Verlag Lond

Vaughan JG (1979) Food Microscopy. Acad. Press, Lond

Wilson T, Sheppard C (1984) Theory and Practice of Scanning Optical Microscopy (Acad Press NY)

Chapter 9

Freezing of Fruit and Vegetables

B. W. W. Grout, G. J. Morris and M. R. McLellan

Introduction

Many soft fruits are harvested and eaten at a developmental stage (ripeness) which is at, or close to, the onset of senescence. The consumer recognises a particular texture for these fruits when they are eaten which is largely derived from the turgor within the tissue. This element of turgor is due to cellular compartments having fully functional, semi-permeable limiting membranes, and retaining solutions with a significantly negative water potential. The turgor generated within the cells resists the mechanical stresses of bite until its yield point, when there is a dramatic collapse, (Brown 1977, Reid 1987). Mechanical properties of cell walls and modified cells within the tissue will also contribute to this impression (Szesesniak and Smith 1969; Brown 1979).

The goal for a frozen product is to retain as much of this turgor as is possible, and to approach the "living" characteristics of the fruit such that the consumer will make a favourable comparison with fresh material. Obviously less successful is any frozen product that compares favourably with other preserved, but not fresh, materials. In this situation, therefore, a successful freezing process is one that maintains the osmotic integrity of a maximum number of cellular compartments in the fruit. This should be achieved without compromising any other required properties and efficiently retaining the turgor of the cells. A similar situation exists for salad vegetables that are eaten without cooking, and successful freezing will again rely largely upon preservation of turgor.

This relationship between texture and turgor is less apparent in vegetables usually eaten after cooking e.g. new potatoes, where structural components such as cell walls and enclosed starch grains generate many of the properties perceived as normal texture. The cooking process will destroy the semi-permeable properties essential to maintain turgor, and bulk water will be able to diffuse simply along gradients within the tissues. Much of the texture that is perceived as part of cooked fruits and vegetables will be due to modified and thickened cell walls (Brown 1977). This aspect of texture is relatively robust, but the mechanical properties of cell walls may be significantly reduced by excessive heating, which is why overcooking produces mushiness. The loss of semi-permeability is a loss of compartmentalisation within the tissues. Undesirable chemical reactions may occur between compounds usually kept separate, both during freezing and subsequent storage (Joslyn 1966). Conventionally, various techniques of blanching are employed to inactivate these compounds prior to freezing to minimise the problem. If more of the viable characteristics of the vegetables were preserved by the freezing process, particularly the integrity of intracellular compartments, then these unwanted reactions might be significantly reduced during storage and the efficiency of blanching enhanced (Burnette 1977; Singh and Wang 1977). Similar benefits might also be felt in some of the soft fruit systems included in the discussion above. The required texture for frozen soft fruit is that which brings the material as close to the fresh condition as possible. Where cooking is involved the thawed product must not be indistinguishable from fresh material that has also been cooked.

When attempting to preserve cellular viability by freezing (cryopreservation) it is essential that the processes are designed with due regard to the integrity of cellular compartments. The aim should be to preserve their osmotic status, semi-permeability and discreet chemical inventory together with overall function. Experience gained in critical aspects of process design might be valuable when attempting to improve the freezing of the more delicate types of fruit and vegetable systems. The following discussion enlarges on this approach and shows the benefits of manipulating parts of the cooling/freezing cycle, primarily in fruit tissues, with respect to the retention of turgor and, therefore, texture.

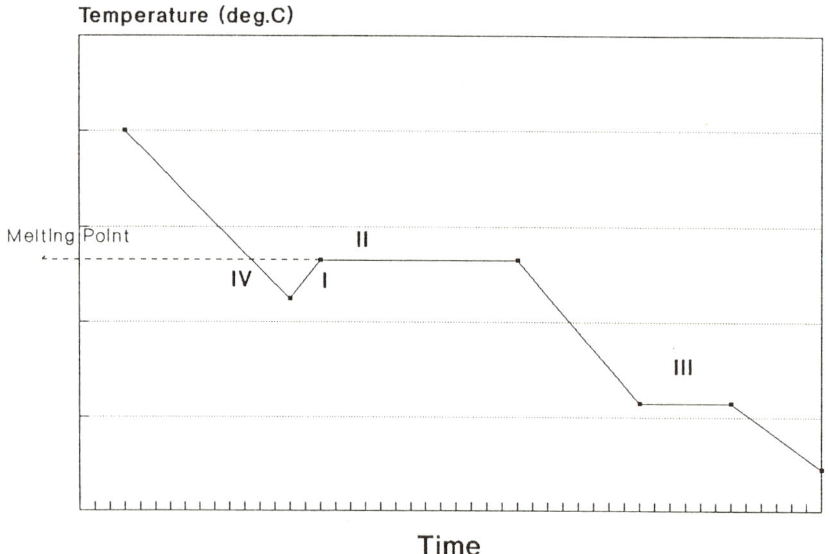

Fig. 9.1. A diagrammatic representation of the freezing processes in dilute, aqueous solutions.
I The exotherm due to heat production as ice first nucleates
IIThe region where most of the latent heat resulting from ice formation is dissipated
IIIEutectic temperature
IVThe undercooled region between the melting point and the point of ice nucleation

Critical Aspects of the Freezing Process

A generalised thermal history for cooled soft fruit tissues is illustrated in Figure 9.1. The curve indicates progressive temperature reduction with an element of undercooling that depends upon both cooling rate and the inherent ice nucleation properties of the tissues (Grout and Morris 1987). In plum slices, for example, several degrees of undercooling are typical at a cooling rate of 8°C per min between +10°C and -10°C, whereas with grape the ice nucleation within the tissues is so efficient that it is often difficult to detect any significant undercooling.

At constant, relatively slow rates of cooling the heat output resulting from the freezing of water in the sample (the exotherm) in typical fruit and vegetable tissues will be large enough to delay significantly further cooling of the tissue (Fig. 9.1). There are two contributing factors operating in this excursion of temperature and time from the required cooling rate. These are:

i) a reduced cooling rate due to the release of latent heat of fusion following freezing of solutions in the material;
ii) a reduced cooling rate due to warming and thermal insulation of as yet unfrozen parts of the sample by the heat output from (i).

Significant delays during processing between the temperature of first-formed ice and the storage temperature will extend the period when the sample is closely associated with ice/solution mixtures. The altered physical and chemical environment in this situation will generate a complex of stresses each of which may be potentially damaging to the component elements of the sample (see Morris 1981; IIR 1986; Grout and Morris 1987).

Because of the large size of the extracellular compartment in living fruit and vegetable tissues it is most probable that ice will form first at this location (Hobbs 1974), and will propagate into the cytoplasm when the plasmalemma loses its semi-permeability due to freezing injury. At this stage other, smaller compartments may still be unfrozen and, in the case of the vacuole bounded by tonoplast, may be sufficiently large to contribute turgor to the texture of the thawed material. The presence of intracellular ice will be lethal, and will result in a loss of compartmentalisation within the tissues. Such material cannot retain water and so loses much of its turgor. Further, the risk of spoilage reactions from the mixing of usually separated cellular components is greatly increased. At some point the smaller cell compartments such as vacuoles and organelles will also contain ice.

The categories of stress associated with extracellular freezing of solutions can be summarised as follows:

(a) Reduction of temperature;
(b) Mechanical effects of extracellular ice crystals including their growth by recrystallisation;
(c) Altered physical properties of the residual, extracellular solution (e.g. viscosity, pH) and changes in molecular/ionic concentration;
(d) Generation of gas bubbles and electrical fields at the ice-residual solution interface;
(e) Concentration of solutes in the extracellular solution and the related decrease in the water potential;
(f) Volume reduction of the protoplast and possible surface area reduction of the plasmalemma;

(g) Increase in concentration of cytoplasmic solutes;
(h) Effects of an increasingly concentrated cytoplasm on the contained organelles.

During freezing the tisses will be exposed to a sequence of potentially damaging stresses, and each of them may have a different importance depending upon the subject of the study (Morris 1981; Grout and Morris 1987). Given the relatively slow rates of cooling that are likely to occur in processing of bulk tissue, and the stabilisation temperature of -20°C, it is likely that ice crystal size will be a significant factor in the injury to most fruit and vegetable tissues (Love 1966; Anon and Calvelo 1980; Reid 1987). At relatively high sub-zero temperatures i.e. following the exotherm, the conditions will be right for the generation of large ice crystals (Franks 1985). Ice crystals will cause disruption and physical lesions in hydrated materials as they invade space previously occupied by disperse solutions, for they will occupy a larger volume than the original space. At any given temperature above the eutectic point, crystalline ice will form and drive solutes into a non-frozen volume, where the accumulating concentration of solute will prevent ice nucleation. As cooling continues more ice will form, reducing the volume of this residual solution and increasing its solute content. The physical properties of the residual solution will be much altered from those of the original, unfrozen sample. High salt concentrations and levels of specific ions, for example, might be particularly damaging to a wide range of materials typical of foodstuffs, particularly if exposure were to be prolonged. A further effect of the changing properties of these solutions will be the generation of electrical perturbance at the interface between the advancing ice front and the residual solution. If the perturbation contacts susceptible structures, such as cellular membranes in fresh fruit tissue, serious injury might result. This will cause loss of semipermeable properties and a consequent reduction in texture and quality of the finished product.

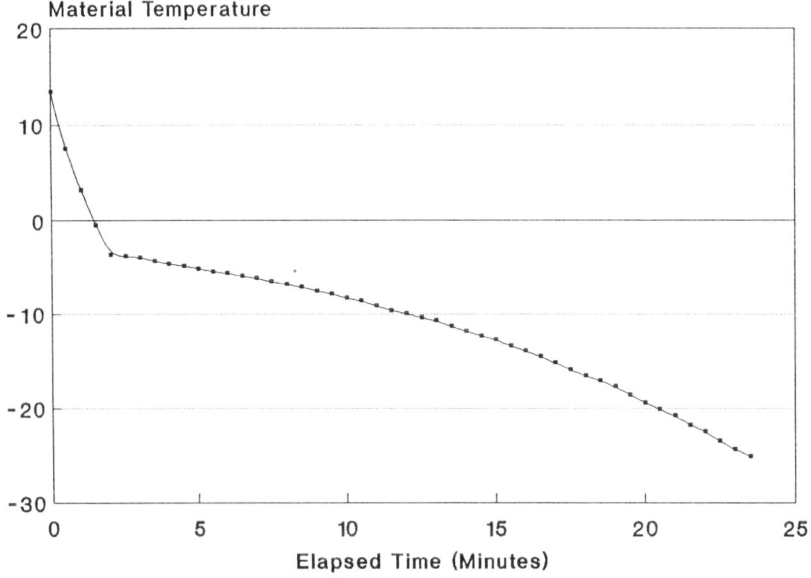

Fig. 9.2. The duration of the freezing process for plum slices coooled in a commercial blast freezer operating at -40°C.

Small gas bubbles generated during freezing as a result of changing solubilities at low temperature will, commonly, travel through the sample causing significant mechanical injury in those systems where fine structure contributes to texture. Hypertonic stresses due to the composition of the residual solution can also compromise the osmotic integrity of viable cells in fresh fruit and tissues, leading to loss of turgor and texture in the product. Further, the free movement of cellular content following such injury will produce the unwanted mixing of hitherto compartmentalised materials. These can give rise to odours and off-flavours during storage.

At a storage temperature of -20°C injuries will still accumulate due to the growth of ice crystals and the effects of any remaining residual solution, providing limits to the practical storage period at such a temperature (IIR 1986). As continued freezing stress, and therefore deterioration in quality of a thawed product, is an inevitable consequence of extended storage, it is important to minimise the levels of stress imposed during processing. This can be achieved, at least in part, by minimising the freezing stage of processing and the time taken to reach the storage temperature. For relatively large tissue pieces eg. small whole fruits such as strawberry, or sliced tissues such as peaches and courgettes, conventional freeze-processing in a blast freezer operating at -40°C will generate a freezing plateau that effectively exposes the tissues to freezing stresses for a time that can be measured in minutes (See Fig. 9.2).

Clearly, this prolonged exposure to the stresses associated with freezing, be it extracellular or within any other of the major cell compartments, will significantly increase the possibility of membrane damage, loss of semi-permeability and, consequently, loss of turgor following thawing. Rupture of tissue compartments, however caused, will cause unregulated loss of solutions, leading to a flaccid recovered product (Szezesnik and Smith 1969; Brown 1979; Mohr 1971, 1974). Additionally, the potential for an increase in unwanted chemical interactions due to random mixing of cellular solutions (see above) will be greater as the plateau associated with the freezing stage of the process is extended.

There is an obvious benefit, therefore, from minimising this plateau for frozen fruits and vegetables, provided it can be implemented in a cost-effective way and without drastic departures from current industial practice. The benefits of such a system, in terms of turgor retention and perceived texture of thawed product, can be illustrated using strawberries as a model. This data, together with the appropriate thermal profiles, is given in the following discussion.

Evaluation

Much of the empirical work concerned with reduction of the freezing plateau was carried out using a microcomputer-based model. This model takes into account:

i) physical properties of the tissue such as size, surface area and volume;
ii) water content and osmolality of contained solutions;
iii) initial and final temperatures;
iv) surface heat transfer coefficient and thermal conductiviy of the product;
v) enthalpy changes;
vi) performance parameters of the cooling equipment.

Although a simple, first-order model it is able to generate data sufficiently close to experimental observations to justify its use (Fig. 9.3).

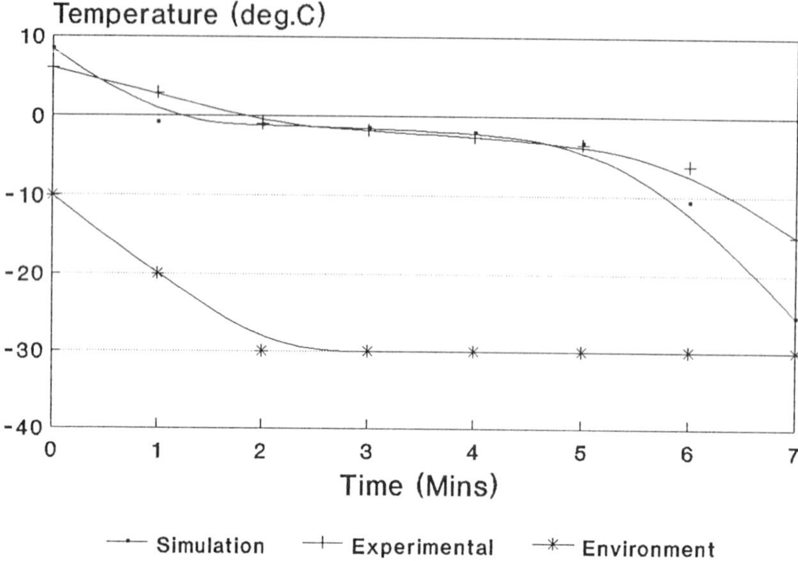

Fig. 9.3. The performance of a computer method used to simulate the temperature response of the tissues during cooling. In this example data for peach slices is compared with the output from the model.

Strawberries were selected at a point where there was a limited amount of unripe surface tissue i.e. yellow/pink not red. This material was shop-bought in the UK in April, and was of Spanish origin. The very limited under-ripeness meant that the strawberries were typically graded as Class 2. The strawberries were halved immediately before freezing and sample temperatures recorded from a copper/constantan thermocouple embedded in the tissues. The material was cooled in a Kryo 10-16 programmable freezer (Planer Products, Sunbury-on-Thames, Middx) and transferred directly from -20°C to a conventional domestic chest freezer also maintained at -20 ± 2°C for storage. Comparisons were made with strawberries frozen by immersion in liquid nitrogen and then returned to -20°C, and with material that had been placed directly into a blast-freezing chamber operating at -40°C. The condition of the thawed product was assessed by recording drip loss and by placing the material before a panel for sensory evaluation. Using the computer simulation of cooling to design a protocol and avoiding environmental temperatures below -60°C during processing, it was possible to reduce the freezing plateau for strawberry significantly (see Fig. 9.4).

The material from each treatment was thawed to room temperature from a storage temperature of -20°C after a storage period of 14 days. When examined some 45 min after reaching room temperature the material frozen with a reduced freezing plateau appeared firm, with little solution lost. This material together with its "conventionally" frozen counterpart were presented to a sensory evaluation panel, together with some material that had been quenched directly into liquid nitrogen. Also included in the evaluation were commercially available frozen strawberries purchased from a local freezer store. The material frozen with a reduced latent heat plateau was rated significantly higher than conventional samples for both texture and flavour, but was not significantly different from strawberries immersed directly in liquid nitrogen (Table 9.1).

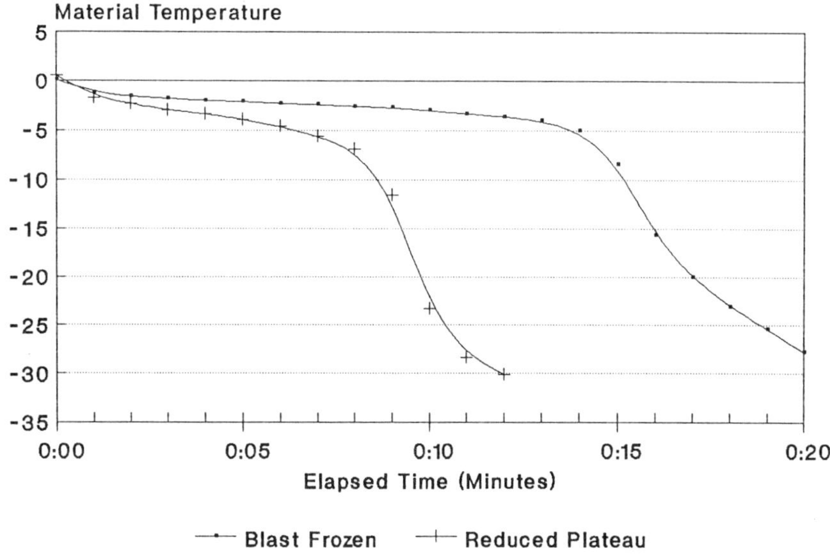

Fig. 9.4. The selected thermal histories for strawberry halves frozen with and without a reduction of the time taken to complete the freezing stage of the process.

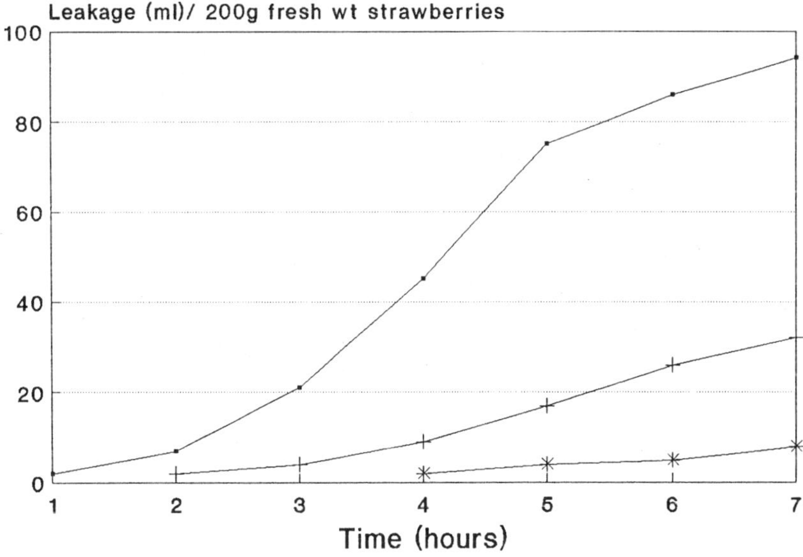

Fig. 9.5. Leakage from thawed strawberry halves following various freezing techniques: A Blast frozen; B Immersed in liquid; C Frozen with a reduced latent heat plateau (See Fig. 9. 4).

As a direct result of the rapid rate at which it was cooled, the material quenched in liquid nitrogen would be expected to have the smallest ice crystals of all the samples and possibly, therefore, the least injury. However, recrystallisation and ice crystal growth during storage would diminish this benefit significantly (Reid 1987). An examination of drip loss from the material in this investigation clearly shows the benefits of the reduced freezing plateau over the liquid nitrogen rapid freeze by providing information on the amount of solution retained in semipermeable compartments within the sample (Fig. 9.5).

Table 9.1. Texture and firmness of 45 min-thawed strawberry tissues n=25

	BLAST	SLH	IN	COMM
Firmness/texture	3.3	3.8	3.7	2.2
	(1.2)	(1.2)	(1.0)	(1.1)
Flavour	2.8	3.2	3.4	2.2
	(1.3)	(1.4)	(1.6)	(1.1)

SLH: Shortened freezing plateau
LN: Quenched in liquid nitrogen
COMM: Commercially frozen material

The delay of drip from the material processed with a shortened freezing plateau to some 4 h after thawing indicates preservation of a high degree of compartmentalisation. This would be expected to give improved turgor when compared to the other experimental material, and to have better storage characteristics in the frozen state. The exudate from the strawberries frozen with the reduced plateau was also paler in colour than that from the other treatments, indicating that it contained less of the vacuolar pigments, which further supports a hypothesis of enhanced preservation of compartmentalisation. A third piece of supportive evidence is provided by the data from experiments where thawing was accelerated by immersing the strawberries in water at 50°C (Table 9.2).

Table 9.2. Exudate loss from frozen strawberries thawed at two rates

Time	Slow thaw*		Fast thaw**	
	A	B	A	B
1 h	0	0	43	6
2	0	0	75	34
3	0.5	0	81	41
4	4	0	82	49
5	19.5	2	83	51
6	41	4.5	87	55

Data collected as ml of exudate for 200g of strawberries subsequently frozen.
* unassisted on bench ** water bath at 50°C
A longer freezing plateau
B shortened freezing plateau

Under rapid thawing conditions a relatively high drip loss was seen in material frozen with the shortened latent heat plateau, but this was drastically reduced when slow thawing was employed as water that had translocated to extracellular spaces had time to move back into cellular compartments across intact semi-permeable membranes.

Summary

The simple demonstration reported above reinforces the concept of turgor as a property of intact, osmotically-functional cellular compartments. Preservation of texture in frozen-thawed soft fruits will, therefore, be dependent upon successful frozen preservation of a significant proportion of these compartments.

The freezing stresses that damage cells such that they lose semi-permeability are predominant in the period immediately following ice nucleation, when the evolved latent heat has to be dispersed. The importance of minimising this period is an essential consideration in viable cryopreservation, and it is clear that there is benefit in applying this approach to frozen fruits that are to be eaten without further preparation.

An increase in the proportion of intact compartments will reduce unwanted mixing of cellular chemical reactants and thus will influence blanching treatments. A freezing protocol that reduces the latent heat plateau may increase the efficiency of blanching, or allow the process to be reduced in some way.

During thawing water will be lost relatively rapidly from halved strawberries by following a pathway of least resistance, which will be through the extracellular matrix. In tissues where a high proportion of the semi-permeable compartments have been damaged the bulk of the tissue solution will be able to drain through this pathway. If, however, a significantly higher proportion of cellular compartments retain semi-permeability water lost during cryodehydration will be able to diffuse back into the cells. Relatively slow thawing and release of water in extracellular regions, will facilitate this process. Rapid thawing, however, releases a large volume of water in a short period of time and facilitates drainage through the extracellular pathway. The replacement of water back into the semi-permeable conmpartments will restore an element of the original turgor of the tissue, with an accompanying improvement in texture.

References

Anon MC and Calvelo A (1980) Freezing rate effects on the drip loss of frozen beef. Meat Science 4: 1-14

Brown MS (1977) Texture of frozen fruits and vegetables. Journal of Texture Studies 7: 391-404

Brown MS (1979) Frozen fruits and vegetables: their chemistry, physics and cryobiology. Advances in Food Research 25: 181-235

Burnette FS (1977) Peroxidase and its relationships to food flavour and quality. Journal of Food Science 42: 1-6

Franks F (1985) Biophysics and Biochemistry at low Temperatures, CUP, Cambridge pp 210

Grout BWW and Morris GJ (1987) Freezing and cellular organisation. In Grout BWW and Morris GJ (eds) The Effects of Low Temperatures on Biological Systems, Edward Arnold, London, pp 147-173

Hobbs PV (1974) Ice Physics. Clarendon Press, Oxford, pp 837

IIR (1986) Recommendations for the processing and handling of frozen foods. International Institute of Refrigeration, Paris, pp 418

Joslyn MA (1966) The freezing of fruits and vegetable. In: Meryman, H.T (ed) Cryobiology, Academic Press, London, pp 565-607

Love RM (1966) Freezing of animal tissue. In: Meryman HT (ed) Cryobiology, Academic Press, London, pp 317-405

Mohr WP (1971) Freeze thaw damage to protoplasmic structure in high moisture edible plant tissue. Journal of Texture Studies 2: 316-27

Mohr WP (1974) Freeze-thaw and blanch damage to vegetable ultrastructure. Journal of Texture Studies 5: 13-27

Morris GJ (1981) Cryopreservation. ITE, London pp 27

Reid DS (1987) The freezing of food tissues. In: Grout BWW and GJ Morris (eds) The Effects of Low Temperatures on Biological Systems, Edward Arnold, London pp 478-488

Singh RP and Wang CY (1977) Quality of frozen foods - a review. Journal of Food Process Engineering 1: 97-127

Szczesniak AS and Smith BJ (1969) Observation on strawberry texture, a three pronged approach. Journal of Texture Studies 1: 65-89

Chapter 10

Physio-Chemical Problems Associated with Fish Freezing

J. Lavety

Introduction

Freezing, cold storage and thawing are, at first sight, well understood processes that are now used as part of the processing chain handling many types of fish and fish products. Freezing has the singular advantage that it is possible to store highly perishable commodities for considerable lengths of time, so that on thawing, it will be virtually indistinguishable in sensory terms and nutrition from the starting material.

Unfortunately, for a variety of reasons, frozen fish in commerce rarely match this ideal.

Scope

This paper describes the freezing, cold storage and thawing of fish and the associated changes that can occur in appearance, odour, texture and flavour.

Definitions

Various terms are used when describing aspects of freezing, frozen storage and thawing. These are often loosely defined but the meanings with reference to this text are as follows:

Chillroom - a storage room or container designed to maintain a temperature above 0°C and below ambient. The usual range for fish and fish products is between 1°C and 5°C depending on the type of product and length of storage.

Cold store - a storage room or container designed to store frozen food. The temperature will depend on the specific use but a stable temperature of -30°C is required to permit a reasonable storage time without serious deterioration for most fish and fish products.

Connective tissue - (myocomma, myocommata), the collagen based tissue that holds the blocks of muscle (myotomes) together. When fish is cooked, the collagen is readily converted to gelatin at temperatures well below the boiling point of water whereupon the connective tissue forms a gel. Hence the tenderness of cooked fish tissue compared to mammalian tissue, where the differently structured collagen may require considerable time at 100°C to be converted to gelatin. The skin of all fish and animals contains large quantities of collagen.

Denaturation - an alteration to protein in which intermolecular cross-linking takes place that permanently changes the physical and chemical properties of the protein.

Drip - the expressible fluid from fish flesh. The volume of drip increases dramatically with denaturation.

Fatty fish - fish such as mackerel and herring that usually contain a considerable quantity of stored lipids. Conversely, fish such as cod and haddock are frequently referred to as non-fatty or lean fish although about 1% of fat is normally present mostly as phospholipid.

Freezer burn - describes the areas of a frozen product in which the ice has sublimed leaving the product dehydrated, porous and spongy. In severe cases, the phenomenon can affect the entire surface and penetrate deeply into the product.

Gaping - The phenomenon exhibited by the cut surface of uncooked fillets in which the connective tissue fails to hold the muscle blocks together. The surface presents a split or cracked appearance and, in severe cases, the fillet will disintegrate if skinned.

Tempering - raising the temperature of frozen fish or product to between -5°C and -10°C to permit a mechanical process such as slicing. Alternatively, fish or products may be frozen to this temperature range for the same purpose.

Background

Although investigations into the cellular effects of freezing started some 70 years ago (Plank *et al.* 1916), followed shortly afterwards by Clark and Almy (1920), and later by Reay (1933) who demonstrated protein changes linked to cold storage. It was Notevarp and Heen (1938) who recognised that the 'drip' or watery fluid that can be squeezed from denatured fish was related to the eating quality. Banks (1955) devised an apparatus to express the drip under constant conditions and was able to relate the volume of drip to the cooked toughness under some conditions although biological and other variations produced unpredictable variability in the results.

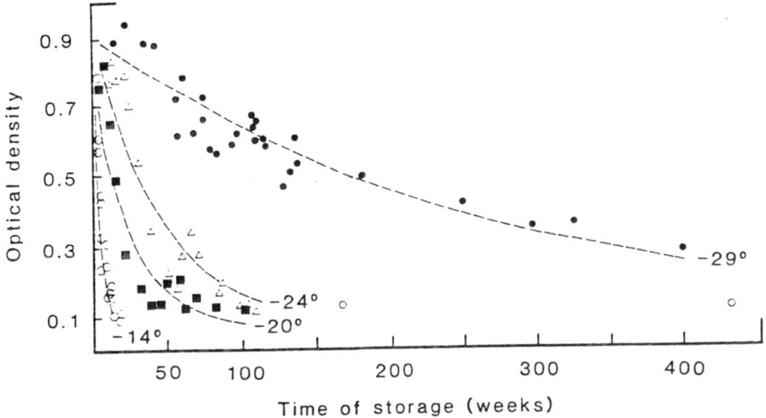

Fig. 10.1. The development of toughness, as measured by the decrease in the optical density of a homogenate, in the muscle of cod stored at different sub-zero temperatures. After (Love 1962a). (Source: The Food Fishes, R M Love, Farrand Press, London (1988))

Ironside and Love (1958) investigated the usefulness of measuring extractable protein as a loss of quality or denaturation indicator and clarified the specific structural changes during frozen storage occurring in the muscle cells as well as the influence of the connective tissue. The mechanisms of the protein denaturation was investigated in detail by Love (1958-1962) culminating in the production of the Torry Cell Fragility Meter (Love and Mackay 1962). This device measured the optical transmission of a muscle suspension and gave results that correlated well with sensory methods of assessing denaturation. The method was later improved by altering the suspension medium (Love and Muslemuddin 1972), the methodology and the apparatus (Whittle 1973 and 1975). Figure 10.1 shows the effect of storage time and temperature on the optical density in the muscle of cod.

In addition to the textural changes, flavour changes occur during frozen storage. The predominant smell and taste that develops in badly cold-stored fish has been likened to that of wet cardboard. McGill *et al.* (1974) succeeded in isolating and identifying the principal component, cis-4-heptenal. The changes that produce this and other compounds are complex but their production is linked concurrently with denaturation as a result of the use of unsuitably high cold store temperature and/or protracted storage.

Freezing

Slow and Fast Freezing

When fish is cooled below about -1.5°C, the water in the musculature will start to freeze. If a high differential in temperature exists, freezing will be rapid and small ice crystals will result. Under laboratory conditions, pieces of fish muscle frozen almost instantaneously will produce ice crystals too small to be observed under a microscope (Menz and Luyet, 1961) whereas if freezing takes about 20 hours, very large ice crystals of up to 10 mm in length can be seen (Lavety 1970 - Unpublished). In the latter case, the fillet has a dark, translucent, vitreous appearance, whereas, the rapidly frozen fillets are densely white and opaque. Properly cold stored, both fillets will thaw to give a satisfactory product.

Long freezing times (> 6 h) are not advocated; in general, shorter freezing times are beneficial in minimising textural and spoilage changes. On the other hand, extremely short freezing times are likely to cause structural damage if the thickness of the fish or fillet exceeds about 10 mm.

Freezer Burn

Inefficient air blast freezers can cause freezer burn (dehydration) but this is rare although up to 1% weight loss due to dehydration may be expected during good conditions in air blast freezing. Unfortunately, the phenomenon is still a serious problem during frozen storage (q.v.).

Freezers

Some typical mechanical designs are outlined in the following text but the emphasis is on the advantages and disadvantages with respect to product quality.

Primary freezing

n this case, the product is in contact with the refrigerant in the form of a cryogenic gas or liquid. Under controlled conditions, excellent results are achieved and the process is particularly suitable for small and/or thin products. These systems produce very little dehydration but may induce considerable thermal stress. If whole fish are frozen particularly rapidly, the exterior freezes to form an extremely hard case which then cracks as the underlying tissue subsequently freezes and expands. Whole fish frozen by immersion in liquid nitrogen will display severe radial cracking with the result that the thawed product will be impossible to process due to the shattered structure.

Secondary freezing

This category includes the very common horizontal and vertical plate freezers where the refrigerant removes the heat by circulation within the plates that are in contact with the product. A number of different refrigerants can be used. The main disadvantage is that irregular shapes are difficult to freeze as good contact with the plates is required for efficient use.

Tertiary freezing

Probably the most significant category of freezer in use today and can be found in a wide variety of shape and form. Essentially, the refrigerant is used to cool air which is in turn circulated over the product (air blast freezing). This category of freezer is particularly suitable for freezing varied and irregular shapes such as large whole fish. Careful design will minimise dehydration (freezer burn).

Gaping

Although principally caused by bad handling, freezing almost always causes some degree of gaping. Fish that display no gaping will gape to some extent after freezing and thawing; refreezing will further advance the condition. This is marked with well-fed fish and is particularly noticeable with round fish such as cod. Flat fish rarely gape to any extent and the increase after freezing is often unnoticeable. Gaping is often mistaken as a sign of spoilage or poor condition. Normal spoilage will have little effect on gaping and, since gaping progressively increases with decreasing water content, gaping can be taken as a sign of relatively high protein content and good biological condition.

Research at Torry Research Station has shown that the best results are obtained if whole fish is frozen pre-rigor. Fish will freeze successfully in *rigor mortis*, but any attempt to straighten out or bend the fish while in *rigor mortis* will cause gaping.

Frozen Storage

Denaturation

Fish protein and lipid are relatively vulnerable components and readily undergo marked changes in the frozen state unless certain precautions are taken.

Fish protein is particularly sensitive to denaturation where the protein develops cross-links between adjacent protein molecules (Jarenback and Liljemark 1975) that effectively stop the thawed fish protein re-absorbing the water to re-create the pre-frozen gel structure. This denatured protein has a much tougher texture than undenatured protein.

Denaturation of the protein will only occur if the duration of frozen storage and the temperature of the cold-store are unsuitable. If the fish is stored for an appropriately short time and at a sufficiently low temperature, the subsequently thawed fish will rehydrate with the protein returning to its original gel condition.

Oxidation

In addition to protein denaturation, other deteriorative processes occur with fatty fish such as herring and mackerel. The rate of oxidation of the relatively reactive or labile unsaturated fats to produce rancidity normally outpaces denaturation and other changes. This oxidative rancidity can be prevented or greatly reduced by glazing or packing to exclude air. Glazing is carried out by quickly immersing the frozen fish in very cold water immediately after freezing which produces a protective shell of ice. Alternatively, vacuum packing in oxygen impermeable plastic bags is used with great success. Whole herring can be frozen and stored successfully by placing the unfrozen fish in a polythene lined paper bag in a vertical plate freezer and topping up the bag with water to fill the voids before applying the refrigerant.

Although fatty fish produce characteristic rancid flavours, a slower oxidative change producing the characteristic 'cold-store' odour and flavour takes place in the low fat white fish such as cod. This is well described by McGill *et al.* (1974) and Hardy *et al.* (1979) with some recent developments in cold store taints described by Josephson and Lindsay (1987) and Karahadian and Lindsay (1989).

Freezer Burn

When the ice in the skin or tissue of the fish sublimes, the dried protein remaining is extensively and irrevocably denatured. This denaturation is the reason why it is difficult to freeze dry fish effectively.

Dehydration or freezer burn is frequently seen in unwrapped, unglazed fish after a term in a cold-store. Some degree of freezer burn is acceptable in 'round' fish (i.e. not filleted) as the skin is normally discarded before consumption, but even a trace on the cut surface of a fillet is likely to lead to rejection. Dehydration is prevented by glazing, packing or wrapping as described earlier in the prevention of oxidation.

Temperature Fluctuation

When fish is frozen, a certain amount of water will remain unfrozen due to the bonds that link it with components such as protein. In addition, salts, enzymes and other constituents that were previously dissolved in the now frozen water are concentrated in any remaining water and assist in preventing it from freezing. When the temperature fluctuates, the dynamic changes accelerate denaturation and the ice crystal size will tend to increase causing further physical damage.

Parasites

The potentially pathogenic parasitic worms that are occasionally found in fish are killed by freezing. Some agencies now require freezing as a prerequisite before purchase or sale of fish.

Thawing

Principles

Thawing is the application of heat to the frozen fish until all the ice present has reverted to water.

In freezing, the heat being transferred to the refrigerant is conducted through tissue that has been frozen. In thawing, the reverse is the case and the heat has to be conducted through thawed material. Since the conductivity of heat through ice is greater than water, the thawing process is likely to take at least three times as long as freezing. In addition, unlike freezing, a large temperature differential cannot be used if spoilage or, in extreme cases, cooking has to be avoided. This is further complicated when thawing whole fish. The tail portion is much thinner than the head end, and will thaw out faster. As soon as this happens, the fish must be chilled to prevent bacterial and enzymatic spoilage. To date, and despite the research activities of many workers, no universally acceptable thawer has yet been designed for whole fish. A common practice is to expose the fish to ambient temperatures until the frost that forms on the surface starts to thaw at the tail end, and then transfer the fish to a chill room overnight. With very large fish, this presents considerable difficulties.

Incomplete thawing, where ice crystals are still present in the deep tissue, is liable to produce gaping and structural damage on filleting unless exceptional care is taken to avoid flexing the fillet. Particular attention has to be given to ensure complete thawing if conveyor belt handling of the fillet is likely.

Thawers

Simple thawing solutions work well in some cases. If circumstances permit, a spray of water at 5°C to 8°C will effectively thaw whole fish. Alternatively the fish can be immersed in a bath of water but problems with cost, hygiene, circulation and effluent disposal exist. A commercially successful vacuum thawer has been developed to thaw blocks of medium and small whole fish. The blocks rest on bars in the thawer and as the thawing progresses, the blocks break up and fall between the bars thus allowing the heat to penetrate throughout the fish. Various other systems using vacuum, high humidity, infra-red and electrical resistance have been found effective for many applications but have had very limited success with fish and fish products. A system using acoustic vibration to assist heat transfer has had conflicting reports. Microwave radiation is used successfully to temper fish but attempts to thaw completely tend to lead to thermal runaway with localised heating.

Discussion

The relative difficulty in freezing and cold storing fish stems from the variation in size, shape and intrinsic condition of the material. Mature fish undergo an annual cycle that usually involves starvation and hence condition varies. For instance, replete cod have a water content of about 80% whereas post-spawned starved cod may approach 86%. The annual cyclic changes in fatty fish can be dramatic with a change in fat content of up to 20%. These and other intrinsic biological variations alter the rate of deterioration during frozen storage.

The relative difficulty in thawing is mainly due to the ease of cooking and the rapidity of spoilage compounded by shape and non-uniformity.

Despite these problems, solutions have been found to effectively freeze and cold store virtually all fish and fish products. By knowing the problems and the characteristics of the specific material, an effective freezing, cold storage and thawing regime can be set up.

Due consideration must be given to each of the three aspects of freezing fish; freezing, frozen storage and thawing. Too often, freezing is given undue importance relative to cold storage and thawing is only considered when it is realised that a problem exists. A holistic approach is required for each application so that freezing is carried out at the optimum point on the processing line. For instance, it may be better to freeze whole fish to minimise the risk of dehydration and accept the thawing difficulties, whereas, if the end product is to be sold to the consumer in the frozen state, every effort should be made to get processing completed without an extra intermediate frozen state. Freezing at sea was once thought to be the only way to get commercially worthwhile quantities of fish from very distant waters to the UK. Now that the problems in freezing and thawing whole fish are evident, fully processing the fish on board the fishing vessel to the stage of frozen consumer-ready packs may be an alternative, providing hygiene and safety criteria can be met.

To achieve good results using freezing, it is absolutely essential that:

1 the temperature of the cold store is sufficiently low to keep denaturation to an acceptable level for the storage period required;
2 the temperature fluctuations are minimised;
3 protection against dehydration and oxygen entry is adequate.

It is also recommended that fish is chilled in ice until a core temperature of near 0°C is achieved before freezing is attempted and that the core temperature is at least down to -25°C before removal to the cold store. Attempts to freeze by directly placing unfrozen or partially frozen fish directly into a cold store will give inferior results and is almost certain to raise the temperature of the contents of the cold store to some degree.

Outlook

The scarcity of fish means that more distant waters are being fished. Consumers are demanding higher quality and commercial and industrial standards are increasing. Freezing is almost the only universally accepted preservation method that does not alter the character of the material. Likely advances will be in more efficient systems with improved packaging and, hopefully, multiple freezing will be reduced.

Acknowledgement

The work described is part of the UK Ministry of Agriculture, Fisheries and Food programme.

References

Banks A (1955) The expressible fluid of fish fillets. II-Method of determination. J Sci Food Agric 6:282-286

Clark ED, Almy LH (1920) A chemical study of frozen fish in storage for long and short periods J Ind Eng Chem 12:656

Hardy R, McGill A, Gunstone F (1979) Lipid and auto-oxidative changes in cold-stored cod. J Sci Fd Agric 30:999-1006

Ironside JIM, Love RM (1958) Studies on protein denaturation of frozen fish, I-Biological factors influencing the amounts of soluble and insoluble protein present in the muscle of North Sea cod. J Sci Food Agric 9: 579

Jarenback L, Liljemark A (1975) Ultrastructure changes during frozen storage of cod (Gadus morhua L) J Food Technol 10:229-239,309-325

Josephson DB, Lindsay RC (1987) Retro-aldol degradation of unsaturated aldehydes. J.A.O.C.S., vol 64, No 1 (Jan)

Karahadian C, Lindsay RC (1989) Evaluation of compounds contribution characterizing fish flavours in fish oils. J.A.O.C.S., vol 66, No.7

Love RM (1958) Studies on protein denaturation in frozen fish. III. The mechanism and site of denaturation at low temperatures. J Sci Food Agric 9:609-617

Love RM (1958) The expressible fluid of fish fillets. IX. Other types of cell damage caused by freezing. J Sci Food Agric 9:262-268

Love RM (1962a) Protein denaturation in frozen fish. VI. Cold-storage studies on cod using the cell fragility method. J Sci Food Agric 13:269-278
Love RM (1962b) Protein denaturation in frozen fish. VII. Effect of the onset and resolution of rigor mortis on denaturation. J Sci Food Agric 13:534-545
Love RM (1962c) New factors involved in the denaturation of frozen cod muscle protein. J Food Sci 27:544-550
Love RM (1966) Bending, shrinkage and texture of cod frozen at sea. J Food Technol 1:137-140
Love RM (1966) The use of tasters for investigating cold storage deterioration in frozen fish. J Food Technol 1:141-146
Love RM (1966) The freezing of animal tissue. In: Meryman HT (ed)"Cryobiology". Academic Press, New York. pp 317-405
Love RM (1966) Protein denaturation in frozen fish. XI. The proportion of tissue water converted to ice. J Sci Food Agric 17:465-471
Love RM (1967) The effect of initial freezing temperature on the behaviour of cod muscle proteins during subsequent storage: a histological study of homogenates. Bull Japan Soc scient Fish 33:746-752
Love RM (1969) Condition of fish and its influence on the quality of the frozen product. In: Kreuzser R (ed) Freezing and irradiation of fish Fishing News (Books) Ltd, London. pp 40-45
Love RM (1969) Time-temperature tolerance. Proc Inst Food Sci. Technol. UK. pp 2,6
Love RM (1969) Anomalous behaviour of frozen cod muscle stored near its melting point. In: Kreuzer, R (ed) "Freezing and irradiation of fish" Fishing News (Books) Ltd London pp 119-127
Love RM, Aref MM, Elerian MK, Ironside JIM, Mackay EM, Varela MG (1965) Protein denaturation in frozen fish. X. Changes in cod muscle in the unfrozen state, with some observations on the principles underlying the cell fragility method. J Sci Food agric 16:259-267
Love RM, Elerian MK (1963) The irreversible loosening of bound water at very low temperatures in cod muscle. CR XI Cong Int Froid 887-892
Love RM, Elerian MK (1964) Protein denaturation in frozen fish. VIII. The temperature of maximum denaturation in cod. J Sci Food agric 15:805-809
Love RM, Haq MA (1970) The connective tissues of fish. IV. Gaping of cod muscle under various conditions of freezing, cold-storage and thawing. J Food Technol 5:249-260
Love RM, Lavety J (1972) The connective tissues of fish. VII. Post-mortem hydration and ice crystal formation in myocommata, and their influence on gaping. J Food Technol 7:431-441
Love RM, Lavety J, Steel PJ (1969) The connective tissues of fish. II. Gaping in commercial species of frozen fish in relation to rigor mortis. J Food Technol
Love RM, Mackay EM (1962) Protein denaturation of frozen fish. V-Development of the cell fragility method for measuring cold store changes in the muscle. J Sci Food Agric 13:200
Love RM, Muslemuddin M (1972) Protein denaturation in frozen fish. XIII-a modified cell fragility method insensitive to the pH of the fish. J Sci Food Agric 23:1239-1251
McGill AS, Hardy R, Burt JR (1974) Hept-cis-4-enal and its contribution to the off-flavour in cold stored Cod. J Sci Food Agric 25:1477-1489
Menz LJ, Luyet BM (1961) An electron microscope study of the distribution of ice in single muscle fibres frozen rapidly. Biodynamica 8:261-294
Notevarp O, Heen E, (1938) The effect of freezing, storage temperature and freshness of raw material on the quality of frozen fish - in Norwegian. Fiskeridir. Skrift., Ser. Teknol. Undersok, 1:No 2, 30, Bergen
Plank R, Ehrenbaum E, Reuter K, (1916) The preservation of fish by freezing. - in German. Z. ges. Kalte-Ind 23:33-75
Reay GA, (1933) The influence of freezing temperature on haddock's muscle. J Soc Chem Ind Lond 52:256T
Whittle KJ (1973) A multiple sampling technique for use with the cell fragility method. J Sci Fd Agric 24:1383-1389
Whittle KJ (1975) Improvement in the Torry-Brown homogenizer for the cell-gragility method. J Fd Technol.10:215-220

Chapter 11

The Special Problems of Freezing Ice Cream

D. W. Everington

Ice cream unlike most other foodstuffs is designed to be eaten in a frozen state and, therefore, ice crystal size influences texture and taste. It is generally accepted that ice crystal size less than approximately 25 μm are undetectable by the palate in the presence of other mix ingredients and give a smooth texture. Larger ice crystal sizes give a coarse texture and therefore control of ice crystal size during the conversion of water into ice during freezing is important in producing ice cream.

Composition

The composition of ice cream mix varies often from country to country, but the composition generally is in the following range:

Composition by Weight	
Fat	7% to 12%
Milk solids, not fat (msnf)	9% to 11.5%
Sugar	12% to 16%
Stabilizer, emulsifier, flavour and colour	0.2% to 1%
Total solids	34% to 40%

The higher the total solids in the mix, the less the amount of water to be converted into ice and even a difference between 36% to 38% total solids has a noticeable effect on texture control.

Water ices generally contain 12% to 16% sugar flavoured with fruit juices, chocolate or artificial flavouring, a stabilizer (alginates, vegetable gums, gelatine, etc) is added to increase the viscosity and assist in the formation of small ice crystals during freezing.

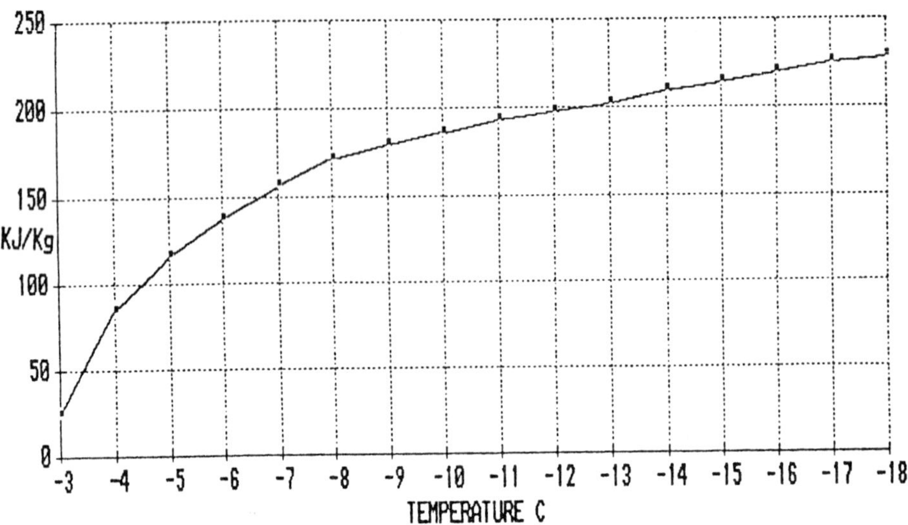

Fig. 11.1. Enthalpy of ice cream as a function of temperature.

Physical Structure of Ice Cream

Ice cream is homogenized by using a high pressure pump discharging through homogenizing valves in order to disperse the fat to a globular size of 0.5 µm to 2 µm. During the initial part of freezing, the air incorporated into the mix called 'over-run' can be varied between 60% to 130%. The frozen ice cream is a foam with air bubbles with a mean diameter in the region of 25 µm, in a continuous phase of sugar syrup in a glassy state containing ice crystals with an approximate mean diameter of 20 µm when measured at -29°C.

Enthalpy of Ice Cream

The enthalpy of ice cream depends on the water content, its initial freezing point and the amount of unfrozen water present when the diameter is reduced to its desired value.

Figure 11.1 shows the enthalpy of ice cream mix calculated by Heldman (1966). The composition of the mix being:

	%
Butterfat	10.1
Solids, not fat	12
Sucrose	14
Corn syrup solids	3.73
Stabilizer	0.22
Total solids:	40.05

Heldman's theoretical approach determines the weight of frozen and unfrozen portions of the mix as a function of temperature, the enthalpy at any temperature below freezing point is then given by

$$H = h_s + h_u + h_l + h_i$$

where: h_s = sensible heat above freezing point
h_u = sensible heat of unfrozen portion
h_l = latent heat of ice
h_i = sensible heat of ice

The calculations were made on the basis of a mix temperature of 4.4°C and this method shows that the sensible heat of frozen and unfrozen portions of the product below its initial freezing point accounts for 9% of the enthalpy when freezing to -18°C.

Freezing of Ice Cream Mix

Freezing takes place in two stages. The initial stage is in a scraped surface heat exchanger, know as an ice cream freezer, in which air is incorporated during freezing to an outlet temperature of between -4°C to -9°C. The second stage of freezing is then normally termed "hardening" where the temperature is reduced to -18°C in an air blast or plate freezer.

Ice Cream Freezer

The ice cream freezer consists of a barrel surrounded on the outside by boiling refrigerant, normally ammonia. A dasher shaft rotates on the inside of the barrel fitted with scraper blades which whips air into the ice cream mix whilst continuously scraping the heat transfer surfaces. Dasher shafts are designated in terms of displacement of cylinder volume, a 15% displacement of cylinder volume produces a wetter texture product desirable for extrusion of shear sensitive products and for fluid products filled into moulds. An 80% displacement dasher shaft is used for the manufacture of low temperature ice cream.

The foam produced by the mixture of air and mix is rapidly frozen in order to produce a large number of ice nuclei, the resultant ice crystal size will be a function of outlet temperature. It can be seen from Figure 11.1 that when the outlet temperature from the freezer is -5°C approximately 50% of the heat has been removed, but when the outlet temperature is lowered to -7.8°C a further 25% of the heat is removed leaving the hardening process to remove only 25% of the heat to freeze to -18°C.

As the mix temperature is lowered, its viscosity increases which increases the dasher pump horsepower. With an outlet temperature of -6.6°C, the heat equivalent of the dasher horsepower equals the heat to be removed from the ice cream. When the temperature is further reduced to -7.8°C, over 70% of the heat load comes from the heat equivalent of the dasher and pump horsepower. It follows, therefore, that as the outlet temperature is reduced the rated capacity of a given heat exchanger surface decreases.

Low temperature ice cream due to its viscosity is difficult to handle and it is this factor which has prevented its widespread use in ice cream manufacture.

Investigators Frazeur and Harrington (1968) have established that low temperature ice cream withstands heat shock better than ice cream frozen at -5°C prior to hardening.

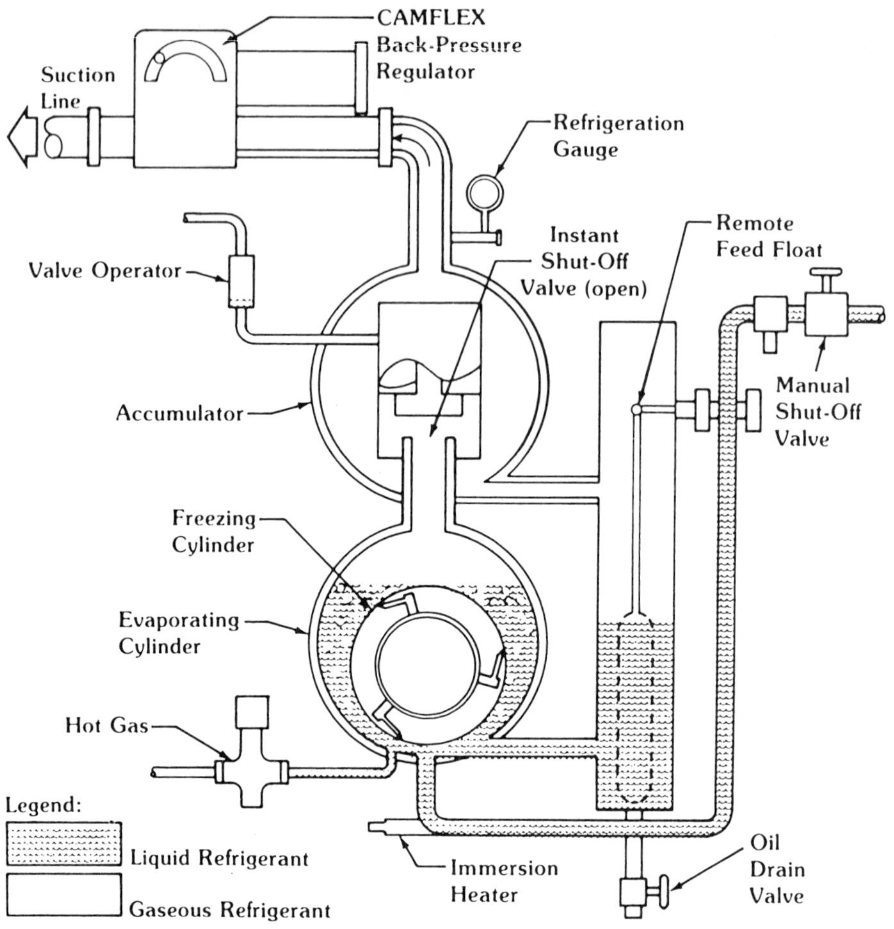

Fig. 11.2. Refrigeration circuit of a flooded ice cream freezer.

Overall heat transfer coefficients obtained are in the region of 2KW/m^2°C when using ammonia with an evaporating temperature of -32°C. Internal coefficients are high due to efficient scraping action and are in the order of 4.5KW/m^2°C while outside boiling refrigerant coefficients are approximately 6.2KW/m^2°C. The presence of oil on the outside surface of the barrel will reduce the boiling refrigerant coefficient by a minimum of 10%, which has led some producers to use oil-free compressors or compressors fitted with very efficient oil separators.

Figure 11.2 shows the refrigeration circuit of a flooded ice cream freezer, together with an instant shut-off valve. If the freezing temperature falls sharply due to abnormal conditions, the viscosity quickly increases causing the dasher shaft power to increase which eventually leads to a freeze-up condition. To prevent this, the instant shut-off valve closes which elevates the evaporating temperature, the valve then opens when normal conditions resume.

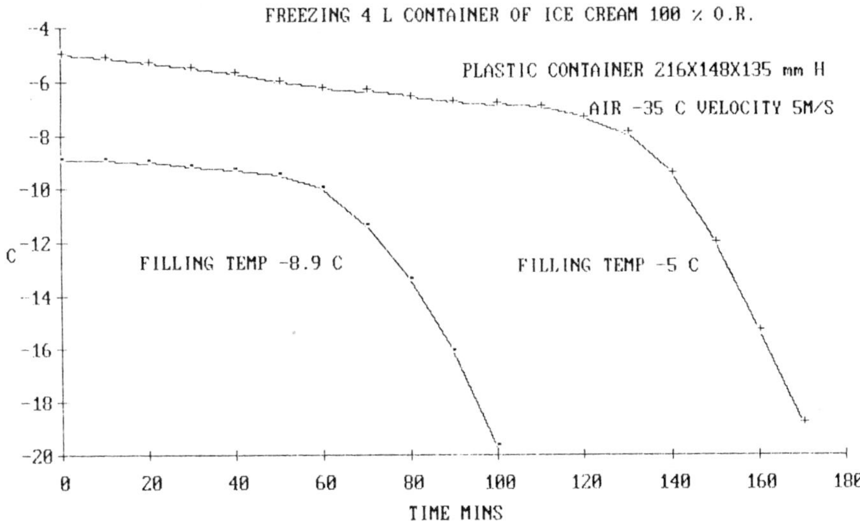

Fig. 11.3. Freezing times for a 4 l container of ice cream frozen in a spiral air blast freezing tunnel.

Ice Cream Hardening

It would appear that the resulting ice crystal size is largely determined by the size of the ice crystal on exit from the ice cream freezer and the final hardening temperature. The speed of hardening within limits does not appear to materially affect the final ice crystal size.

Taste panels have found it difficult to detect any difference between wrapped ice cream hardened in 30 min compared to ice cream hardened in 3 h. However, the size of the hardener is a function of the freezing time and, therefore, it is desirable to freeze as quickly as possible in order to minimise both the capital investment and the space occupied.

Air blast hardening is used for both unwrapped and wrapped products where the ice cream is discharged on to stainless steel belts and passed through an airstream, generally at a temperature of -35°C with air velocities of 5 m/s or greater.

Freezing times of extruded ice cream in the form of ribbons can be decreased by 40% when the extrusion temperature of the ice cream is -8.9°C compared to a normal extrusion temperature of -5°C.

Freezing times obtained in a spiral air blast freezing tunnel for a 4 l container filled at -5°C and -8.9°C are shown in Figure 11.3. It can be seen that the freezing time for the 4 l container is reduced by 40% when the filling temperature is reduced from -5°C to -8.9°C.

In an air blast hardener (Fig. 11.4) the heat equivalent of fan power is approximately 30% of the product enthalpy. Heat transfer coefficients obtained in these tunnels vary beween $15 W/m^2 °C$ and $60 W/m^2 °C$, which is low when compared with heat transfer coefficients obtained from scraped surface heat exchangers which are $2 KW/m^2 °C$. Heat transfer coefficients during hardening can be increased by using automatic plate freezers

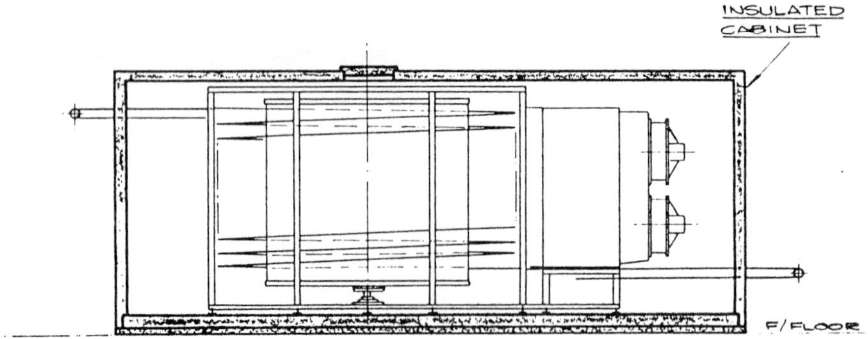

Fig. 11.4. Section through a spiral freezing tunnel.

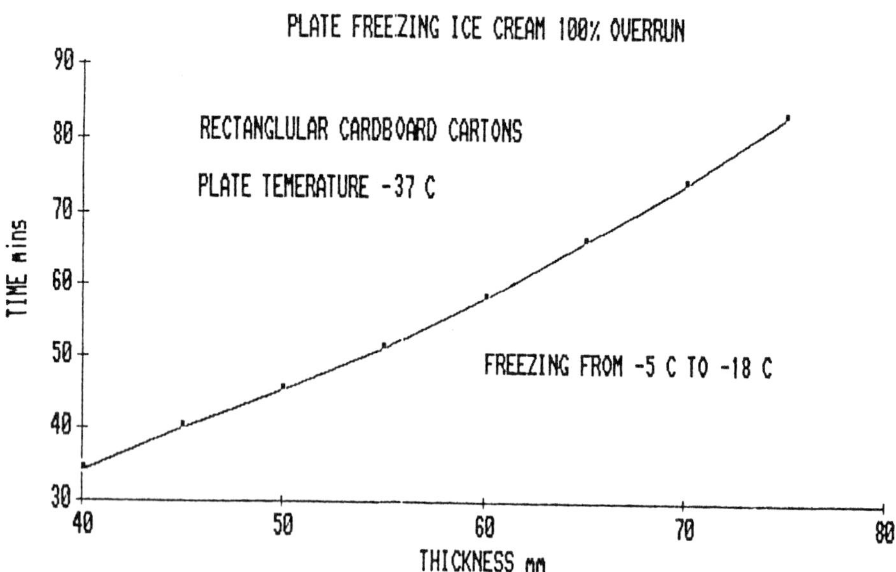

Fig. 11.5. Hardening times of ice cream in cardboard cartons subjected to plate freezing at -37°C.

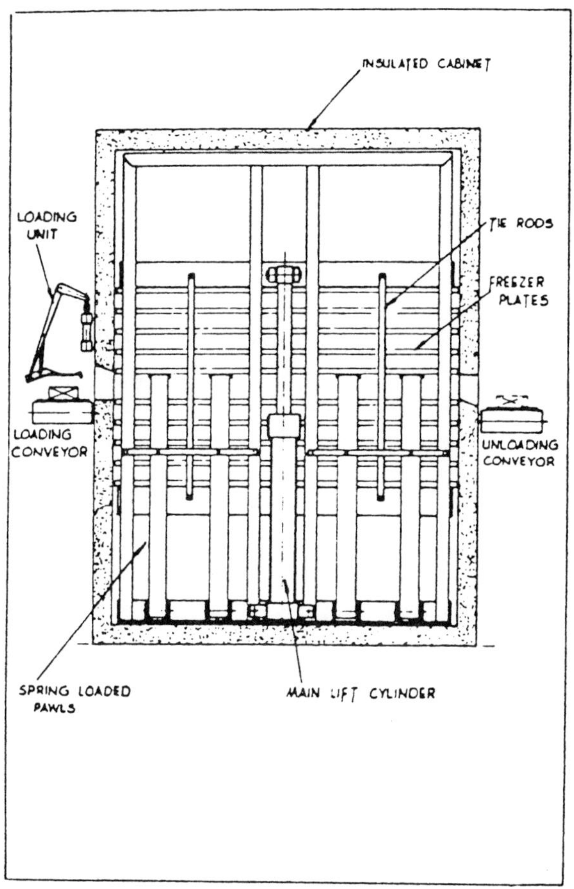

Fig. 11.6. Section through a continuous plate freezer.

where hardening takes place by conduction from refrigerants evaporating in aluminium plates which are brought in direct contact with top and bottom surfaces of the product.

Figure 11.5 shows the hardening times of ice cream filled in cardboard cartons with surface contact area ratios approaching 100% for the top and bottom surfaces.

The advantage of using plate freezers (Fig. 11.6) where possible is that the efficiency of freezing is increased due to the absence of heat input from the fans. As heat transfer coefficients are normally in the region of 500W/m^2°C, hardening times are significantly reduced when compared to air blast hardening. The freezing time for 60 mm thick cartons frozen side-by-side to form an infinite slab in air at -35°C with a velocity of 5 m/s is 120 min, compared to 60 min in a plate freezer with a plate temperature of -37°C. (See Fig. 11.5)

Stick Confections

Ice cream or water ices or a combination of both are frozen in moulds which continuously pass through a brine tank. The brine, normally calcium chloride, is sprayed underneath the mould in order to promote turbulence. The stick confections are released from the moulds by defrosting in a warm brine section of the tank.

When freezing ice cream, a scraped surface heat exchanger can be used to reduce the mix temperature from 4.4°C to near its freezing point. The rapid build-up of viscosity as the mix approaches its freezing point makes it necessary to accurately control the freezing temperature in order to prevent dasher shaft overload. The reduction in freezing time using a scraped surface heat exchanger will amount to 25%, compared to filling the moulds with water ice mix from a storage tank at 4.4°C.

Freezing water ices in an air blast system gives a longer freezing time due to a four-fold decrease in the heat transfer coefficient compared to a brine system. Air freezing water ice therefore gives an undesirable large ice crystal size.

Heat Shock

After hardening, when ice cream is packed at ambient temperature, the outside layers will rise in temperature and then be reduced in cold storage. In the centre of the pallet, the surface temperature can take up to three days to drop 5°C in a cold store with an air temperature of -30°C.

Further temperature cycling can occur in distribution cold stores, transportation and the retail cabinet. The slow freezing which occurs after each temperature rise promotes large ice crystal growth, the size of which will depend on the number of heat shock cycles and the temperature range through which the cycles occur. If cycling occurs below -18°C, then the proportion of ice melted between -30°C and -18°C is approximately 5%. (See Table 11.1) When the temperature rises a further 3°C to -15°C, an additional 3.5% of the ice is melted.

Table 11.1. Percentage water frozen in ice cream 36% total solids

Temp °C	% Water frozen	Temp °C	% Water frozen
- 2.5	0	-18	84.5
- 5	50	-20	86
-10	71	-25	89
-15	81	-30	89.5

While it is commercially acceptable that fluctuations below -18°C will occur, fluctuations above -18°C will have a pronounced affect on texture and in order to avoid fluctuations above this limit, the time/temperature profiles which will occur in the distribution chain must be considered. It can be seen from Figure 11.7 that assuming ice cream is hardened to a centre temperature of -18°C and the outer layers have risen to -18°C during packaging and palletisation, then after five days in a cold store at -28°C the centre temperature drops to -23°C while the outer layers reach cold store temperature.

In an 8 h delivery period between factory and depot cold stores in palletised delivery vehicles at -18°C, the outer layers rise to the vehicles' air temperature. In order to deliver to retail cabinets at -18°C, vehicles making 30 deliveries per day must start the journey with ice cream at an equalised temperature of between -23°C and -24°C which is achieved by holding in a depot cold store for two days at a temperature of -23°C.

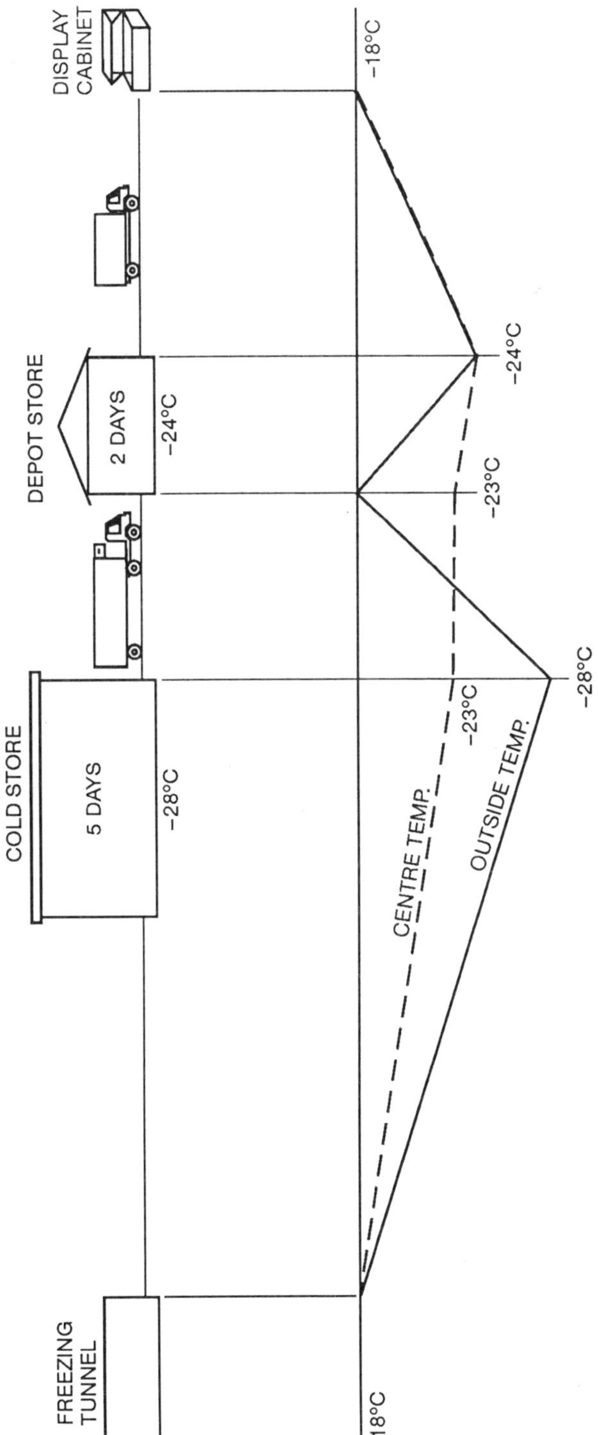

Fig. 11.7. Temperature-time profiles in the distribution chain for ice cream.

It can be seen that if deliveries to retail cabinets take place after hardening to -18°C, then the temperature on delivery is unlikely to be satisfactory. In the particular case considered, the ice cream would need to be hardened to a minimum temperature of -24°C.

In conclusion, ice cream is more susceptible to heat shock than most other frozen foodstuffs. Resistance to heat shock can be achieved by using low temperature extrusion wherever possible and hardening to temperatures which take into account fluctuations which will occur in the cold chain.

Unfortunately, a processor loses control when the ice cream enters either a distribution chain or retail cabinets and the consumer judges quality after further fluctuations have occured both in the retail cabinet and the domestic refrigerator.

References

Heldman DR (1966) Quarterly bulletin on research from Michigan State University 49 (2) 144-54

Frazeur DR and Harrington RB (1968) The effect of storage conditions and heat shocks on body and texture, Food Technology Vol. 22, 910

Keenley PG (1979) Confusion over heat shock, Food Engineering June 1979, Pennsylvania State University

Chapter 12

A Convenience Born of Necessity: The Growth of the Modern Food Freezing Industry

L. Eek

Food Preservation by Freezing

Frozen foods are very close to, and sometimes better than the available fresh counterparts. From a sensory as well as from a nutritional point of view the quality is generally higher than for any other preserved product. Methods of preserving food have become increasingly important. The need for longer shelf life and improved taste and quality have given rise to the modern food freezing industry.

Artificial freezing of foodstuffs started in the second half of the 19th century. One of the first shipments of frozen carcasses of meat, veal and mutton was carried in good condition all the way from Buenos Aires to Rouen, France by the vessel Frigorifique.

Clarence Birdseye - the father of consumer packed frozen products - introduced the first two-sided contact freezer for commercial use in 1929. Consumers were then presented with a whole new set of options for many fruits, vegetables and fish.

The purpose of this paper is to describe the variety of food freezing methods available and to provide a strong basis of information to help food processors decide which system will best suit their needs.

Food Preservation

All food preservation methods are directed to inhibit or decrease the rate of the various reactions responsible for food deterioration. All of these reactions are among other factors influenced by the temperature. The rate of the reactions will decrease at lower temperatures. Cooling and chilled storage, therefore, are well proven ways to enhance the storage life of most food products. But even at temperatures near the freezing point of some reactions, including growth of many microorganisms, continue at a rate which will limit the shelf-life of most food to a relatively short period of time.

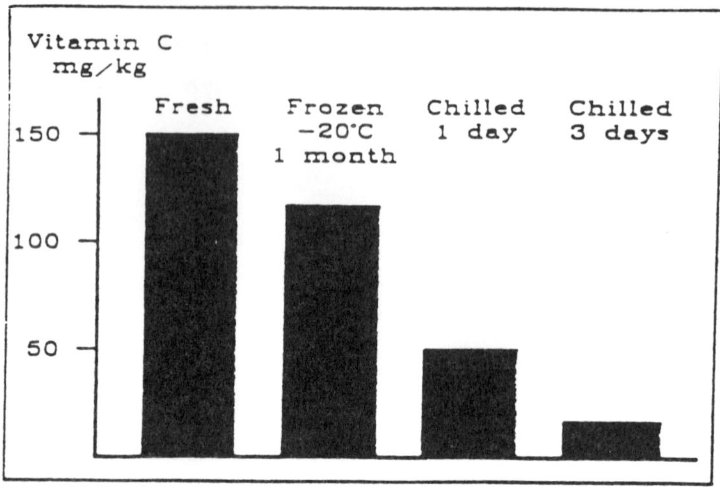

Fig. 12.1. Vitamin content in newly harvested, frozen and chilled peas.

A classical illustration of the difference between a chilled and frozen product is given in Figure 12.1 showing the vitamin C content in newly harvested, frozen and chilled peas.

Various species of microorganisms - bacteria, yeast and mould - are present in all foods prepared or unprepared. Their number depends among other factors on the quality of the raw material and on the hygienic conditions under which the product has been handled. Even if the cooking temperature in preparing a food product is high enough to destroy the organisms present, contamination will nearly always take place. If the product is not handled under strictly hygienic conditions the contamination can be serious. Several kinds of these microorganisms are able to grow and multiply at chilled temperatures and may eventually cause spoilage of the product.

Most organisms present in food are not harmful to man and, even if they cause spoilage of the food, they do not constitute a real threat to the health of the consumer. However, in some cases the food may be contaminated with pathogenic bacteria which, when the food is ingested, can cause illness and food poisoning. Food poisoning can also be caused by metabolic products and toxins from some of these organisms.

The shelf-life of chilled products is enhanced by preservatives, modified atmosphere packing and, possibly in the future, by irradiation.

The demand for fewer additives creates a new problem, that of microbiological hazard - an increased risk for food poisoning. This problem is also relevant to some traditional chilled products which rely on preservation systems based on acids, salt, nitrate and phosphate.

It is obvious that most preservation methods used today will also be used in the future. Freezing has shown itself to be a universal method applicable to raw materials as well as to consumer and catering packed products.

Optimisation of Cooling and Freezing

Today's cooling and freezing equipment is tailored to meet the different products to be handled. It is still true for most products that the faster the freezing, the higher the quality. In reality, however, the rate of freezing is determined by optimising both the quality obtained and the cost of freezing. The cooling process has often been neglected, but is equally important to control, especially for microbiological reasons.

A chilling operation immediately after the heat treatment, meaning that the product is packed when cooled down below 10°C and then frozen, will reduce the possibility of bacteriological growth although some growth can be recorded. If the products, packed or unpacked, are frozen in-line immediately after heat treatment and thereafter packed, there will be almost no increase in the number of bacteria present.

During freezing and frozen storage many microorganisms are injured and even killed. This has, however, a very limited influence on the bacteriological quality of the food. From a practical point of view it can be stated that during frozen storage no microbiological growth will occur.

After thawing there will be a further increase in the microbiological count, starting off at the point where the growth was arrested during the freezing process.

The Freezing Process

The freezing process may be seen as a lowering of the product temperature from its original value to the storage temperature.

From each point within the product the heat must be removed by conduction to the surface. The surface heat is then removed to the refrigeration medium. The freezing time depends on a number of factors, of which the most important are the dimension and shape of the product, the thermal properties, initial and final temperature and, of course, the temperature of the refrigeration medium.

Ice Crystallisation

Most food items consist of, or contain, animal and/or vegetable cells forming biological tissues. The water solution of the tissue is contained between the cells - extracellular fluid - and within the cells - intracellular fluid. The concentration of salts and other solubles is higher within the cells than outside. The cell membrane acts as an osmotic barrier and maintains the difference in concentration.

When the product is frozen the first ice crystals are formed outside the cells since the freezing point is higher than for the more diluted fluid inside. If the freezing rate is low the cell will lose water by diffusion through the membrane and the water will crystallise to ice on the surface of the crystals already formed in the extracellular space. As the cells lose their water the remaining solution within the cells becomes more and more concentrated and their volumes shrink causing the cell walls to collapse. The large ice crystals formed outside the cell wall occupy a larger volume than the corresponding amount of water and therefore will execute a physical pressure on the cell wall. In some cases this pressure can be high enough to damage the cell wall and contribute to an increased drip loss on thawing.

If the freezing rate is high a large number of ice crystallisation nucleii are formed resulting in a much smaller size of the final crystals. However, even in the case of a high freezing rate, the crystals are formed outside the cells. Only at very high freezing rates are small crystals formed uniformly throughout the tissue, both externally and internally with regard to the cell. Such higher freezing rates can only be achieved in comparatively small products.

The importance of the size of the ice crystal has long been regarded as crucial for the quality of the frozen product. It appears from experience that the difference in crystal size and distribution have little affect on the final quality of the thawed product.

However, this does not mean that the freezing rate is negligible; on the contrary freezing time must, in good commercial practice, be determined for each product with regard to:

i microbiological considerations;
ii mechanical losses - from product sticking to a conveyor belt, or product packages dropping to the floor;
iii damage or down grading;
iv dehydration;
v the freezing process must fit into the production line.

Definitions

As has been mentioned earlier, freezing should be carried out as quickly as possible in order to achieve as high a quality of the product as possible. From a technical/operational point of view a more strict definition is needed. Definitions of freezing time, freezing rate and speed of freezing are most useful for comparison of system and equipment. The book "Recommendations for the Processing and Handling of Frozen Foods", 3rd Edition 1986, issued by The International Institute of Refrigeration, 177 boulevard Malesherbes, F-75017 Paris, gives the following definitions:

The Freezing Process

During the freezing process different parts of the product will pass through the various stages at different times. If a particular location in a product is considered, three stages of temperature change can be defined:

Prefreezing Stage The period which elapses between the time at which a product with a warm temperature is subjected to a freezing process and the time at which the water starts to crystallise.

Freezing Stage The period during which the temperature at the considered location is almost constant because the heat which is being abstracted is causing the majority of water to change phase into ice.

Reduction to Storage Temperature The period during which the temperature is reduced from the temperature at which most of the freezable water has been converted to ice to the intended final temperature. The final temperature can, at least, be the storage temperature reached in any part of the product, including the thermal centre or the equilibrium temperature. The equilibrium temperature is the temperature which is achieved under adiabatic conditions - without heat exchange with the environment -

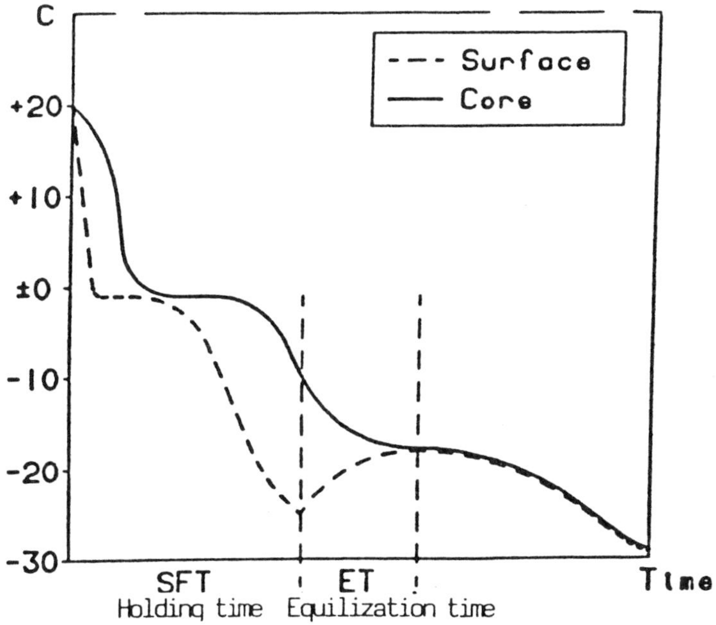

Fig. 12.2. The commercial freezing process.

when there is a close approximation between the surface temperature and the temperature at the thermal centre.

Freezing Time

The freezing time is defined as the time elapsed from the start of the prefreezing stage until the final temperature is reached. The freezing time depends not only on the initial and final temperature of the product and the quantity of heat to be removed, but also on the dimensions (especially the thickness) and shape of the product as well as on the heat transfer process and its temperature.

From a practical point of view the freezing time is defined as the time required to lower the temperature of the product to an equalisation temperature of -18°C under adiabatic conditions. This definition determines the capacity of the freezing equipment. This practical commercial definition of the freezing process is described in Figure 12.2.

The time the product is held in the freezer is known as Standard Freezing Time (SFT) or Holding Time. ET in the figure stands for Equalisation Time. Normally all freezers are designed on this condition. Contrary to what is commonly believed, the temperature is not brought down to -18°C for the entire product in the freezer. This temperature is achieved during handling or packaging prior to storage.

Freezing Equipment

Today's freezing equipment can be divided into two main groups; one integrated in the processing line and one operating in batches, the former being the most predominant.

There are basically three main types of equipment based on the method of heat transfer:

Air blast freezers which use air for heat transfer. Because air is the most common freezing media this method of heat transfer has probably the largest range of designs.

Contact freezers Heat transfer occurs through conduction. A refrigerated surface is placed in direct contact with the product or package to carry away the heat.

Cryogenic freezers use liquid gases, nitrogen, carbon dioxide to produce vapours that pre-cool and freeze the products.

The design of the freezing equipment should optimise the total freezing process. Among the priorities of design criteria the following can be mentioned:

Product quality;
Minimum product losses;
Reliable operation;
Simple operation and maintenance;
Refrigeration economy.

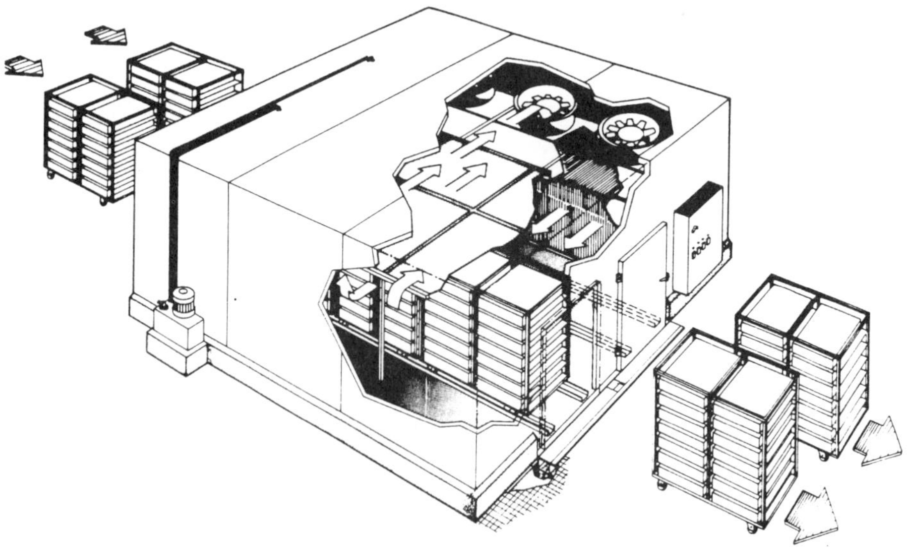

Fig. 12.3 Cut-away section showing movable trolleys in blast freezing tunnel.

The freezing equipment indicated above can also be divided into two distinct groups with regard to the product frozen; individually quick frozen - IQF - and packed products. The advantage of the former are obvious and for these reasons an ever increasing quantity of foods is frozen in this way.

Classic Stationary Tunnel

The classic stationary tunnel, or Blast Tunnel, is the simplest type of freezer which can be designed to produce satisfactory results for the majority of products. It is an insulated enclosure, equipped with refrigeration coils and fans which circulate the air over the product in a controlled way. The product, packed or unpacked, is put on trays which are placed into a rack. Spaces are left between every layer and the racks are positioned in the tunnel in such a way that the air cannot bypass them. The racks are moved in and out of the tunnel manually.

The flexibility of this type of freezer is balanced by high manpower requirements and considerable weight losses if improperly used. Packages containing homogeneous products tend to bulge.

A certain degree of mechanisation is achieved when the racks are fitted with castors or wheels. The racks or trollies are usually moved on rails by a pushing mechanism, often hydraulically powered. (See Fig. 12.3)

This type of freezer generally has the same advantages and disadvantages as the classical tunnel except that it is better suited as an in-line freezer. Labour costs can be reduced and the flexibility is somewhat better as different products can be handled at the same time by different trucks having different dwell-time.

Belt Freezers

The first belt freezers basically consisted of a wire mesh belt conveyor in a blast room which satisfied the need for a continuous product flow. In addition to the disadvantage of poor heat transfer in a blast room many mechanical problems arose.

Modern versions of the belt freezer use a vertical air flow whereby the air is forced through the product layer which creates good contact with all product particles. A condition is, however, that the product is evenly distributed over the whole belt area. Modern designs of belt freezers are mostly based in the spiral belt freezer concept. In these freezers a product belt that can be bent laterally is used. The original spiral belt design utilises a spiralling rail system to carry the belt.

The latest design uses a self-stacking, self-enclosing belt for compactness and improved air flow control. This eliminates the traditional rail system and friction drive. The number of tiers in the belt stack can be varied to accommodate different capacities and line layouts. The belt is continuous which is an advantage as the product transfer points are avoided. These points are the most likely to cause problems with regard to deformation at the beginning of the freezing process and breakage when the product has been crust frozen. The products are placed on the belt outside the freezer where it can be supervised and will stay on the same spot until leaving the freezer. As there is just one belt it is easy to arrange for proper cleaning outside the freezer. A typical design of a modern spiral belt freezer is shown in Figure 12.4.

Both unpacked and packed products are frozen and the freezer gives a large flexibility both with regard to product range and freezing time.

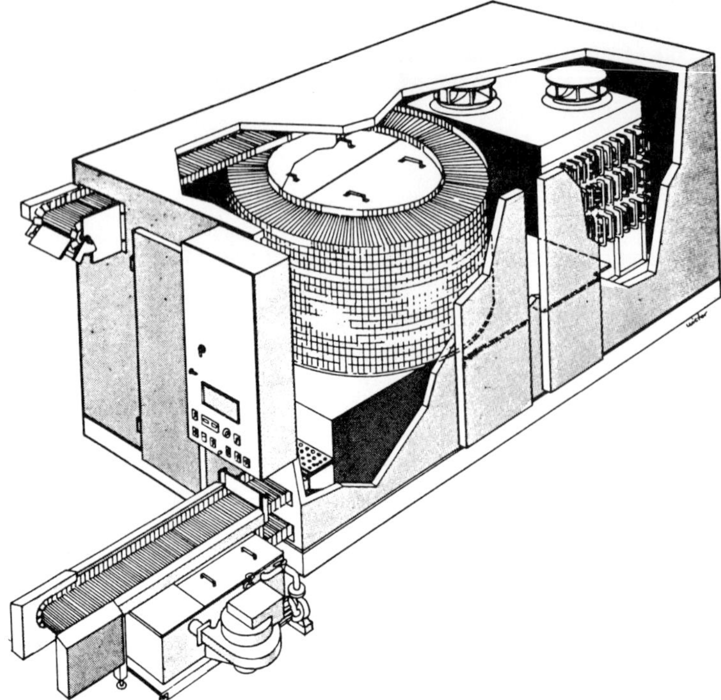

Fig. 12.4. Cut-away section showing a typical modern spiral belt freezer.

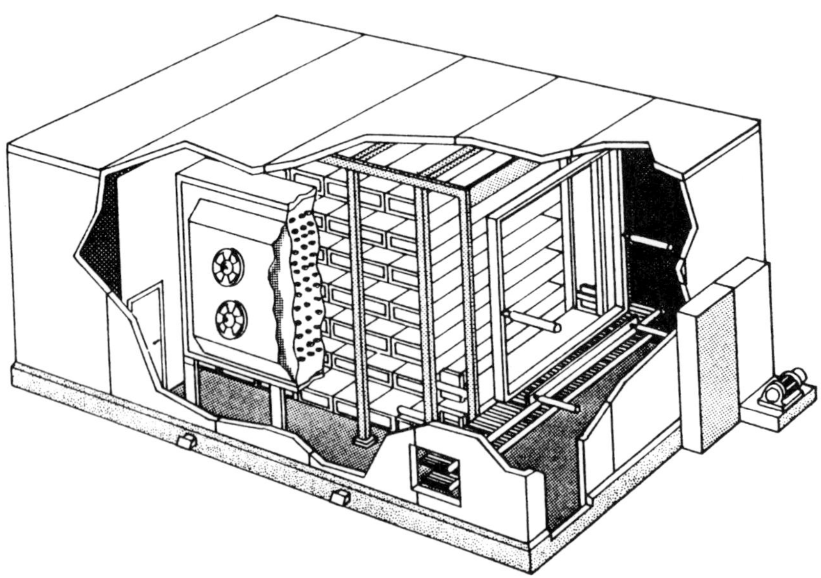

Fig. 12.5. Cut-away section of a carton freezer.

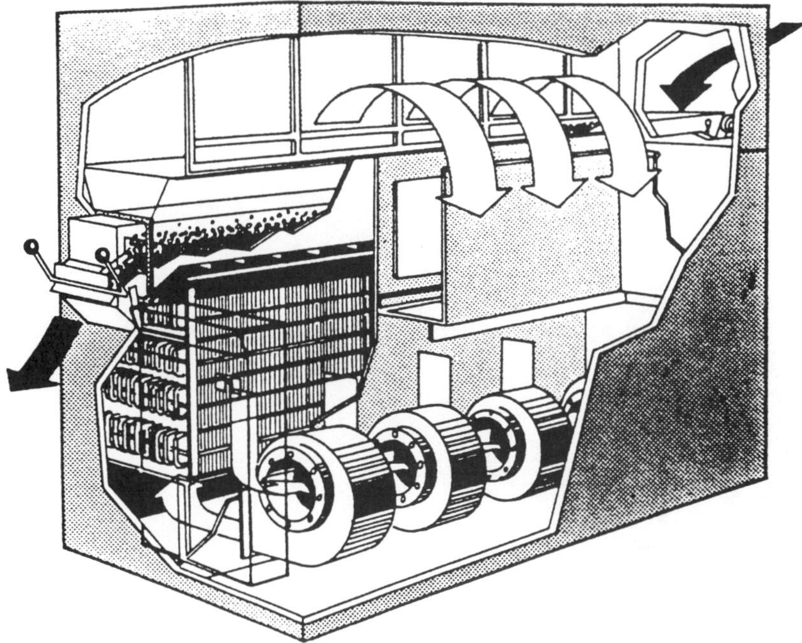

Fig. 12.6. Pictorial view of fluidised bed freezer.

Carton Freezer

This freezer consists of a number of carrier shelves which are automatically moved through the freezing section of the unit. The operations carried out by hydraulic power with mechanical linkage to co-ordinate different movements. The boxes are fed automatically into the freezer on a feeding conveyor.

The carton freezer is used for handling packaged products in larger cartons such as poultry meat, ice cream etc. (See Fig. 12.5)

Fluidised Bed Freezer

This freezer utilises air both as the medium of heat transfer and for transport. The product flows through the freezer on a cushion of cold air which totally surrounds the product. The technique has revolutionised IQF-freezing of fruits, vegetables, shrimps, diced meat and other particular food products. Further, it is ideal for individual quick freezing of products that tend to stick together.

Fluidisation occurs when particles of fairly uniform shape and size are subjected to an upward air stream. At a certain air velocity the particles will float in the air stream, each one separated from the other. In this state the mass of particles can be compared to a fluid. If it is held in a container, which is fed on one end and the other end is lower, the mass - fluid - moves to the lower end as long as more products are added. The typical design is shown in Figure 12.6.

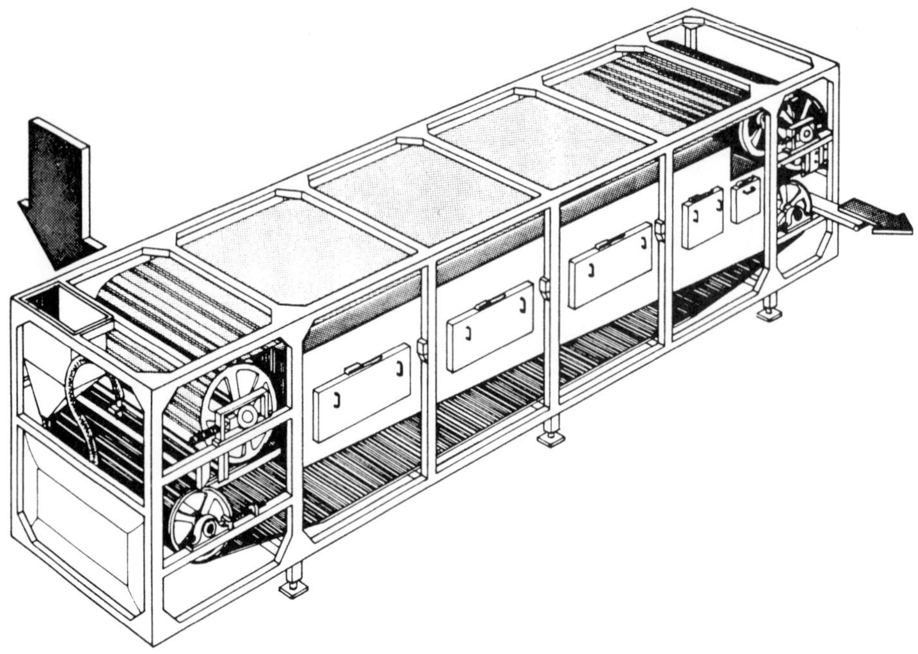

Fig. 12.7. Typical contact or plate freezer.

The use of the fluidisation principle gives a number of advantages in comparison with the use of a belt freezer. The products are always truly individually quick-frozen and this is also true for products with a tendency to stick together. Furthermore, the equipment is totally independent of fluctuations in load. If partly loaded, the air distribution can be the same as for the full load i.e. no risk of the air by-passing the products. The reliability when freezing wet products is greatly improved because the deep fluidised bed can accept products with more surface water.

The fluidised bed freezer can be combined with a belt to a fluidised belt freezer. This freezer is designed to extend the range of IQF products right up to heavyweight items such as corn on the cob.

Contact Freezers

In this type of freezer a refrigerated metal surface is placed in direct contact with the product or package to be frozen. This arrangement gives normally a very good heat transfer which is reflected in short freezing times, provided the product itself is a good heat conductor. The most commonly known representative of this type of equipment is the plate freezer in which the products are placed between refrigerated plates either manually or automatically.

Typical products to be frozen in this type of equipment are different sauces, dips, vegetable purees and fruit pulps. (See Fig. 12.7)

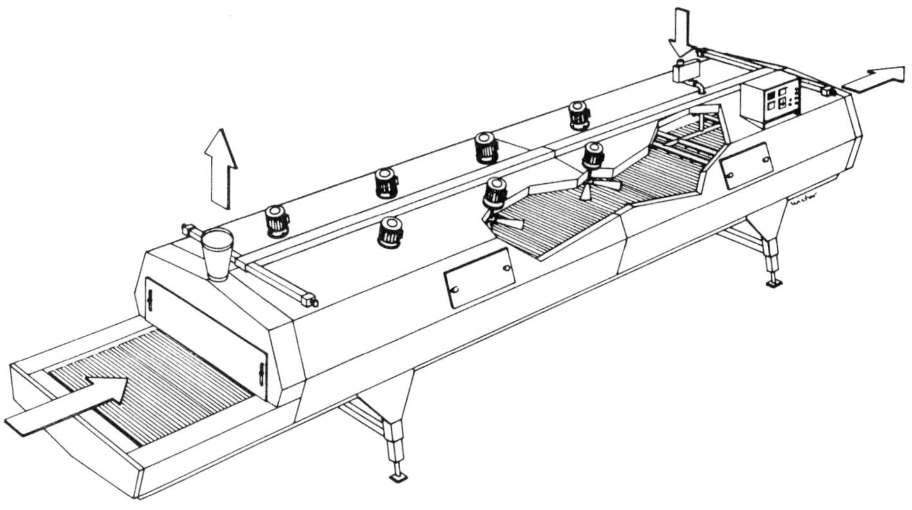

Fig. 12.8. View of straight belt cryogenic freezer.

Cryogenic Freezers

Cryogenic freezers differ from all other freezers in one fundamental way. They are not connected to a refrigeration plant. The heat transfer media are expendable gases that are liquified at large industrial installations and shipped to the food freezing factory in pressure vessels. Two main gases are used - Nitrogen (LIN) and carbon dioxide.

The same basic equipment can be used for both gases. (See Fig. 12.8) Both straight and spiral mesh belts are used. The liquid freezant is sprayed into the enclosure at a suitable location. The freezant evaporates immediately on contact with the products and is allowed to escape after the vapours have been used for pre-cooling of the products. Fan-like "turbulators" create turbulence to improve heat transfer.

The low temperature available, -195°C for LIN and -78°C for carbon dioxide, gives very fast freezing which is a well proven quality advantage for some food products.

The freezant consumption is of the order of 1.0 kg for every 2.0 kg of product to be frozen.

CRUSToFREEZE

A recent development is a combination of the cryogenic freezing system and the air blast system. The equipment utilises the possibility of a fast and efficient crust freezing of extremely wet, sticky products or for handling sensitive products which can then be easily handled in a spiral belt freezer or a fluidised bed freezer without deformation or breakage. This type of freezer offers a possibility to individually freeze products (IQF products) which normally create freezing problems in conventional systems. A CRUSToFREEZE design is shown in Figure 12.9.

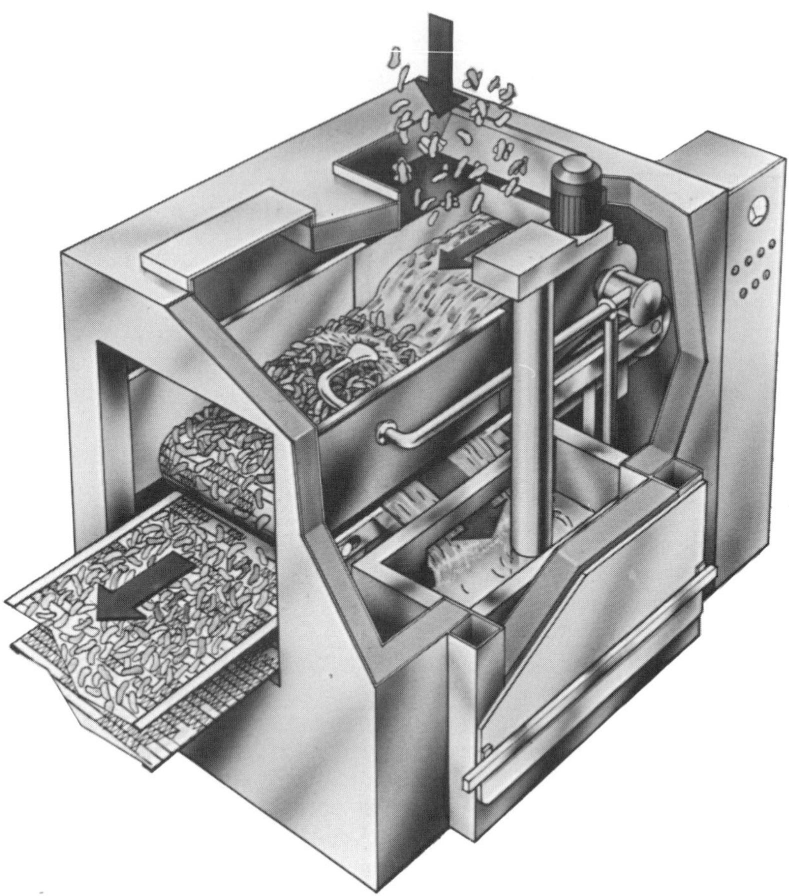

Fig. 12.9. CRUSToFREEZE design which combines the cryogenic and air blast freezing systems.

Selecting Freezing Equipment

Frozen foods are in most cases judged as close to, and even sometimes better than, their available "fresh" counterparts. Both the sensory and nutritional properties are generally better and higher when the food is frozen than treated by any other type of preservation method. The quality of the frozen product depends to a large extent on the freezing method used and the storage time.

The freezer is often the single most expensive part of the processing line. It is important to make sure that the equipment is properly integrated into the process and is working to optimum performance. Careful planning is essential.

With reference to the equipment, operating costs are of obvious importance. High operating costs can offset the advantage of a low initial investment cost. The value of the food products that pass through a freezer in a few weeks time is often many times higher than the investment cost. For that reason reliability is of the utmost importance. More easy to judge when it comes to operating costs are factors like energy consumption, labour and maintenance. The latter calculation should include frequency and degree of ease to maintain the equipment. Downtime and product loss from dehydration are other crucial factors.

Conclusion

The rapid changes taking place in the food supply systems of today call for flexibility. A system that is easy to modify will make it possible to respond quickly to the requirements of the market place.

Today cooling and freezing equipment is tailored to the different products to be handled. It is true for most products that the faster the freezing the higher the quality. In reality, however, the rate of freezing is determined by optimising the quality obtained and the cost of freezing.

Chapter 13

The Use of Liquid Nitrogen in Food Freezing

J. P. Miller

Introduction

Over the past three decades, Europe has witnessed the widespread development of the frozen food industry. Despite the already enormous growth experienced by the industry, the demand for frozen foodstuffs remains increasingly large, fuelled by ever changing lifestyles.

The future growth of the frozen food market will depend on the ability of companies to continue developing freezing and chilling processes which will maintain high standards of product appearance, flavour and nutritive value, whilst ensuring a good return on investment.

Concentration in the industry by the acquisition of many independent companies has seen the growth of automated "assembly line" techniques for the production of sophisticated recipes and products. This in turn has increased the requirement for Individually Quick Frozen (IQF) ingredients of high quality with free flow characteristics to aid portioning, ease of handling and stock control.

Cryogenic food freezing using liquid nitrogen or carbon dioxide as an expendable refrigerant has been established in Europe since the early 1960's. The high pricing of bulk carbon dioxide in the UK has meant that liquid nitrogen has become the most widely used expendable refrigerant in the UK.

Cryogenic freezing offers a range of unique benefits, including:

High cooling rates (quick freezing)
High throughput/low floor space
Flexibility (adaptable to different products, high turn up capability)
Low capital entry

These advantages have driven the growth of the technique so much so, that, an estimated 10% of all frozen food in the UK is now frozen on cryogenic freezers.

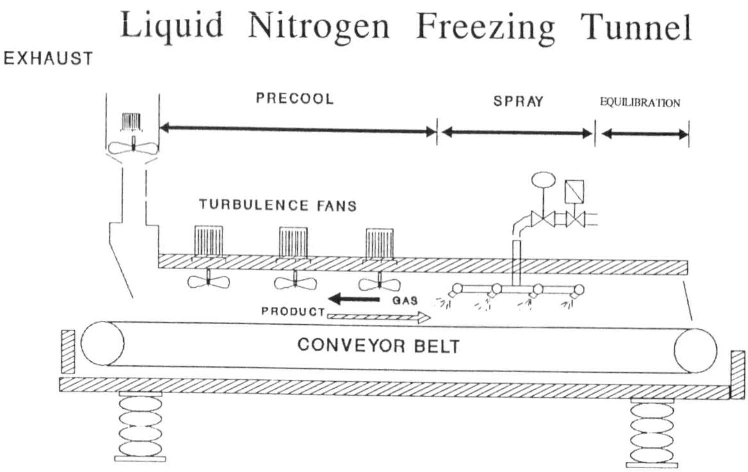

Fig. 13.1. Schematic of cryogenic food freezing tunnel.

Solid and semi-solid foods were the first products to be cryogenically frozen commercially and the first part of this paper describes a range of cryogenic techniques for flowline production of IQF products. The freezing of liquids and suspensions is now an emerging market area, growing rapidly because of the requirement for greater consumer and industrial convenience, the second part of this paper describes systems for the production of individually quick frozen (IQF) liquid pellets.

The Cryogenic Freezing Tunnel

A typical tunnel, shown in Figure 13.1, is essentially a counter-current heat exchanger between the food product and the cryogen. The product is transported on a conveyor belt through a well insulated tunnel most commonly made from stainless steel or other food compatible materials. Overall design must be compatible with food industry standards with particular attention to hygiene and ease of cleaning.

Solid and Semi-Solid Food Freezing

Since 1923 the efforts of Clarence Birdseye and others has ensured the rapid growth of the "Quick-Freezing" industry. The continuous production cryogenic food freezing method, the freezing tunnel, was first introduced in 1962. The flexibility of the tunnel to freeze a wide variety of food products has made it the mainstay of the cryogenic food business and it is still finding wider application within the food industry.

Principles of Operation

The tunnel (see Fig. 13.1) can be split into three cooling regions, the spray zone, the precool zone and the equilibration zone.

Spray Zone In the spray zone, liquid nitrogen is sprayed by a network of atomising nozzles onto the surface of the product close to the product outlet from the tunnel. In absorbing heat from the product surface, liquid nitrogen is vaporised to cold nitrogen gas.

Precool This cold gas from the spray zone is passed counter-current to the product flow precooling the product and being warmed in the process. It is exhausted into the atmosphere at the product inlet. To maximise heat transfer in the gas zone fans are used to agitate the gas and introduce mixing as it passes down the tunnel.

Equilibration Some tunnels incorporate a third short zone after the spray zone to allow conduction of heat from the centre of the food product to the much colder surface crust produced in the spray zone. On exit from the tunnel the equilibration will of course continue in the food package or cold store following freezing.

The major part (60%) of the refrigeration available is obtained from boiling liquid nitrogen at -196°C in the spray zone to nitrogen gas at a similar temperature. The remaining 40% is obtained through heat transfer to the cold gas (HTC).

The heat transfer co-efficients (HTC) in the spray zone are in the range of 100 to 140 $Wm^{-2}K^{-1}$, in the gas zone HTC's are in the range of 40-60 $Wm^{-2}K^{-1}$

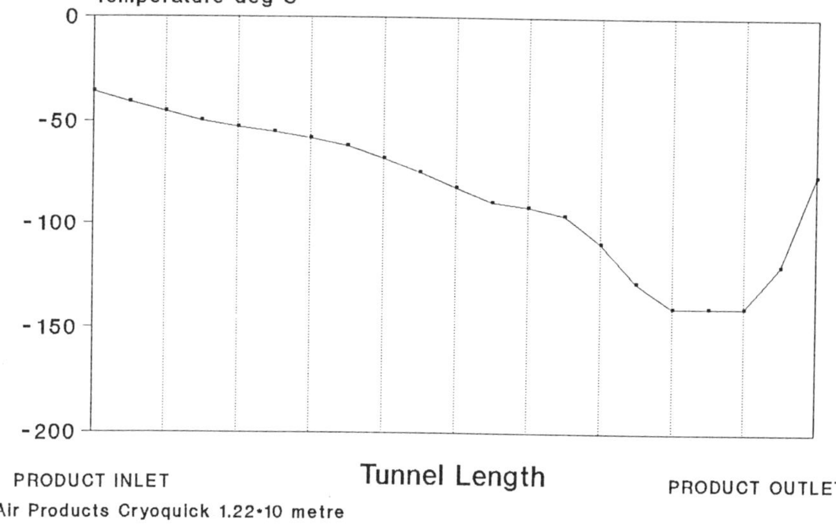

Fig. 13.2. Temperature profile in a liquid nitrogen freezing tunnel.

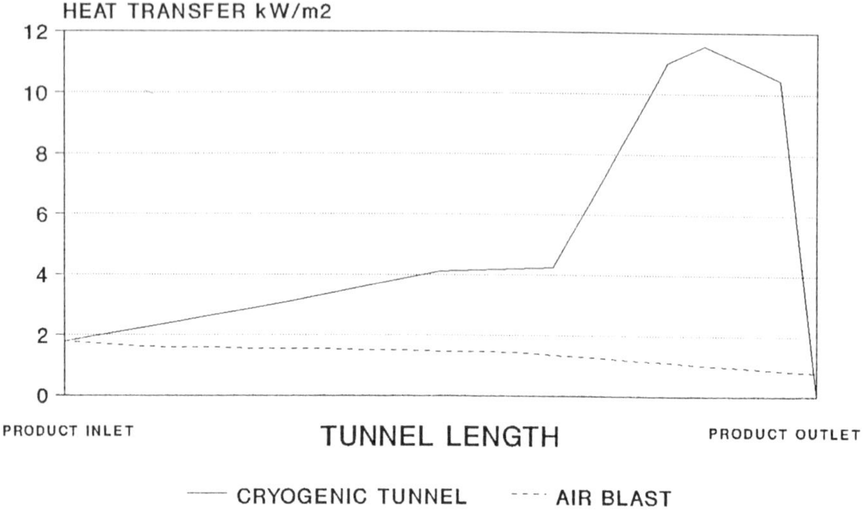

Fig. 13.3. Heat transfer comparison liquid nitrogen versus mechanically refrigerated freezing tunnels.

The combination of these typical heat transfer co-efficients and the high temperature difference between refrigerant and product, produce a heat transfer capability that is typically three to four times greater than that experienced in conventional mechanically refrigerated air blast freezers. Figure 13.2 depicts a typical temperature profile in a cryogenic tunnel freezer.

Figure 13.3 shows the change in heat transfer along the length of a typical liquid nitrogen cryogenic food freezer and compares it with that obtained in a typical air blast tunnel freezer.

Process Control

Final product temperature is a function of the intergrated cooling rate over the final product residence time within the freezer. The cooling is simply controlled by varying the flowrate of liquid nitrogen to the freezing tunnel using a flow control valve. Feedback can be adequately provided by monitoring a relevant temperature within the tunnel. The control of product residence time is accomplished by varying the speed of the conveyor belt. Consequently the throughput is a function of residence time, the belt loading, the width and the length of the tunnel.

Whilst there is no theoretical limit to the size of tunnel freezers, standard machinery is readily available for throughputs ranging from 100 kg per hour to 3000 kg per hour.

Development

Development of cryogenic tunnel technology is continuing both from within the bulk gas supply industry and from external agencies and independent equipment suppliers.

Process development has focused on improvement of convective heat transfer in the gas zone, the automatic control of the temperature profile and efficiency monitoring of the whole process. The goal is to raise overall cooling rates and efficiency thereby reducing the unit consumption of cryogen and increasing product throughputs for a given size freezer.

Equipment development has concentrated on the accessibility and ease of cleaning and improvements in construction techniques to accommodate the large thermal contraction/expansion experienced in the structure from the extreme temperatures. The development is continuing in the 1990's.

Products

When freezing or chilling foods, the ideal products to take advantage of the high cooling rate of a cryogenic tunnel are ones which have a high surface area to volume ratio and where thermal diffusivity of the food does not restrict the rejection of heat to the cryogen. Typical examples include fish fillets, shellfish, pastries, burgers, meat slices, sausages, pizzas, diced and extruded products which are fully frozen.

Alternatively the benefits of cryogenic freezing can be employed to produce a hard crust on a soft product to allow handling or packaging or to aid further processing such as slicing. Typical examples include ice cream, bacon sides, gateaux and whole salmon.

The importance of IQF products has been highlighted in the introduction. Products that are relatively free of surface moisture can be individually quick frozen on a belt freezer, products that are wet at the surface, either naturally or from processing, can present problems such as product "clumping" or stickage to the freezer belt. Examples of such products are fruits, blanched vegetables, raw or cooked shellfish etc. This has driven the development of a new freezing technique "Liquid Nitrogen Immersion Freezing" which is described below.

Immersion Freezing

The freezing of products by simply immersing them in liquid nitrogen has been widely used for many years both in industry and research. The extension of this primarily manual technique to a flow line process of industrial scale has taken many years of development, due to the extremely low temperatures the equipment has to contain and operate within.

The development has been further driven by environmental concern over immersion processes using chlorofluorocarbons (CFC's) particularly RI2. The recirculating CFC processes typically leaked large quantities of RI2 to the atmosphere. This has been as much as 2 tonnes per week of RI2 for a large producer.

Equipment such as "CRYODIP" manufactured by Air Products PLC has found wide acceptance in the freezing of fruits and seafood and has replaced the CFC immersion process in Europe.

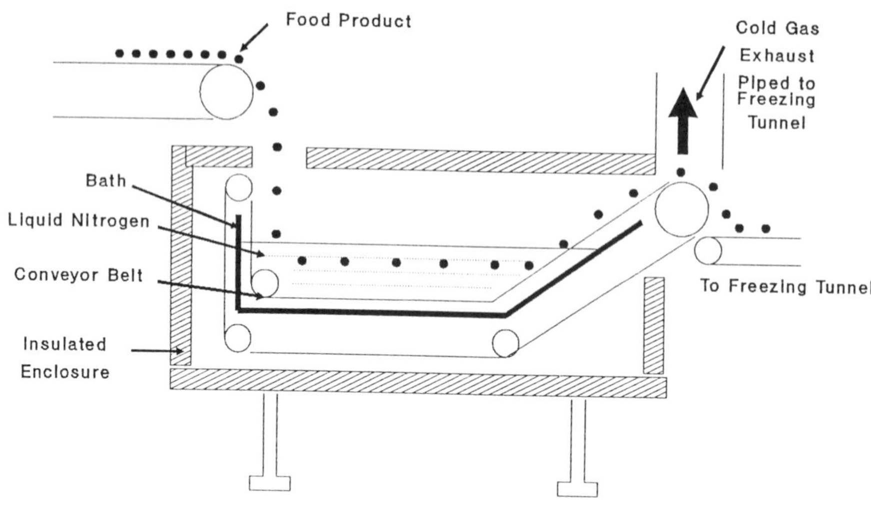

Fig. 13.4. Schematic liquid nitrogen immersion freezing.

Principles of Operation

It is worth listing the key requirements for a high quality IQF process:

Individual frozen pieces without need for further separation;
Minimal product damage from thermal shock through control of cooling rate;
The ability to deal with irregular and varying shapes.

The immersion freezer shown in Figure 13.4 is essentially an efficiently insulated box, fabricated in stainless steel. Food products are conveyed through or allowed to drop into a bath of liquid nitrogen at -196°C, maintained at a constant level. A conveyor lifts the individually crust frozen product from the liquid nitrogen. When the product is fed directly into the bath of liquid nitrogen the crust freezes rapidly in seconds causing the nitrogen to boil violently. This boiling produces extreme turbulence within the liquid nitrogen helping to separate the individual pieces of product while their surface is being frozen. Irregular and complex shapes of food products can be dealt with as the liquid nitrogen accesses all the surfaces of the product.

The residence time in immersion is controlled partly by the speed of the conveyor which lifts the product from the liquid nitrogen bath. Control of residence time is important to prevent "over-freezing" and, with some food products, cracking from thermal shock.

The complete freezing process normally consists of two pieces of equipment, an immersion freezer coupled to a separate belt freezer. The crust frozen product from the immersion freezer is passed directly to the belt tunnel where the freezing process is continued at a slower rate.

The gas produced in the immersion freezer is passed to the belt freezer in the direction of the product flow. The heat exchange is CO-CURRENT. Heat from the product warms the nitrogen gas as it passes along the tunnel to the exhaust. In this manner maximum utilisation is made of the nitrogen refrigerant.

The utilisation of two separate pieces of equipment allows belt speeds to be individually adjusted to provide the requisite heat transfer in each section. Belt speeds in the tunnel freezer can be much slower than the immersion freezer, allowing long residence times with stacking of the product.

Rotary Tunnel IQF Products

A recent development for certain food products that are able to withstand tumbling without damage is the liquid nitrogen rotary tunnel shown in Figure 13.5. The rotary tunnel can achieve high throughput for small areas of floor space with high IQF quality and can offer economies on the usage of liquid nitrogen as described below.

Developed from rotary kiln technology, the process consists of a rotating insulated stainless steel tube which is inclined at a shallow degree to the horizontal. Food products are conveyed into the tunnel and immediately refrigerated by sprays of liquid nitrogen crust freezing the product. The inclination and rotating motion tumbles the product through the drum together with the resultant gas from the boiled liquid nitrogen. Frozen product and warmed nitrogen gas are separated at the exit hood. The constant movement of the product from the tumbling action ensures a high quality IQF product.

As with the immersion process the heat transfer is CO-CURRENT. Whilst the heat transfer coefficient in the convective section of the freezer is lower than fan blown systems, the food product surface area exposure is enhanced by the tumbling action. The overall heat transfer can be very similar to fan agitated systems, but there is now no heat input from the gas turbulence caused by the circulating fans. In this manner more efficient usage of liquid nitrogen refrigerant is achieved with lower consumption of refrigerant pro rata. Typical examples of products successfully frozen in this manner are minced, diced cooked and raw meats, also diced vegetables.

IQF Freezing of Liquids

The freezing of liquids and suspensions is finding greater application throughout the food processing industry to aid stock and or production control or to provide all year round supplies of a seasonable product.

Some liquid products are currently bulk frozen inside plastic containers with typical volumes of between five and 50 litres in conventional air blast freezers. The volume of these containers and the limits of the food products thermal diffusivity prevent rapid freezing. De naturalisation of the product can occur, with release of free water and cold storage life may be affected.

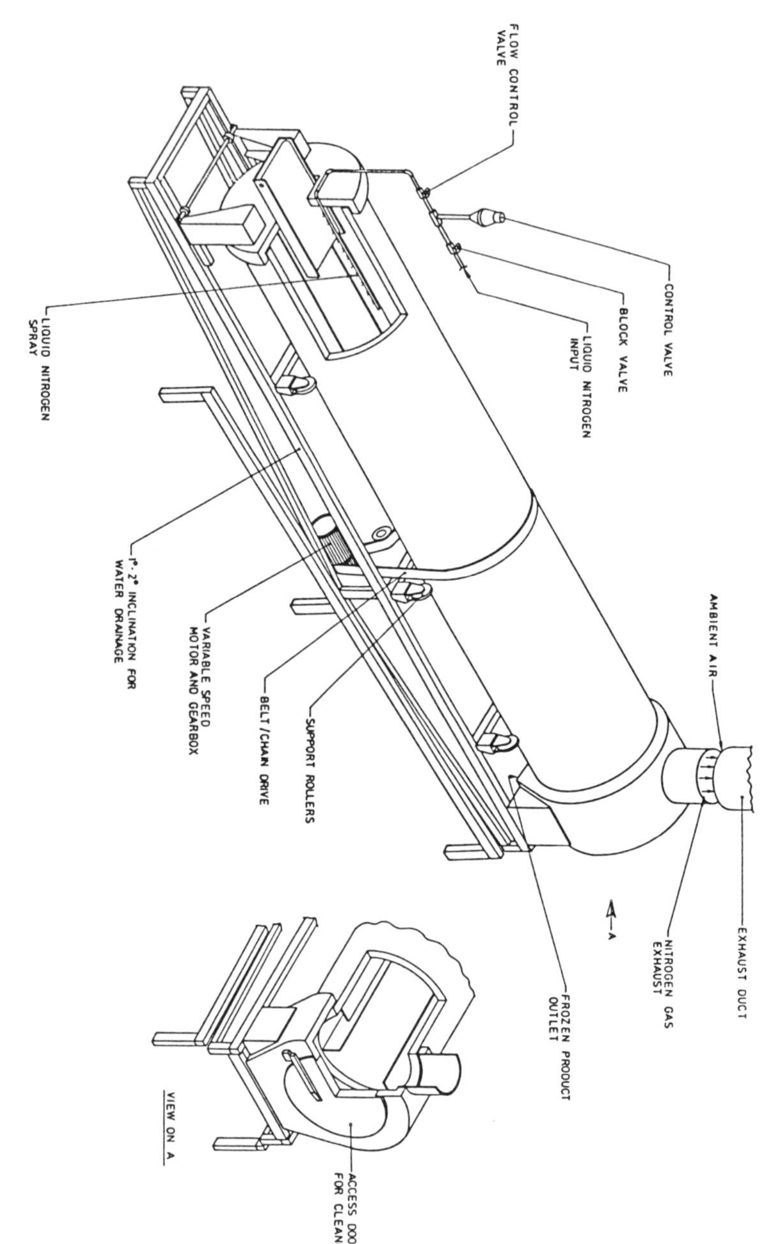

Fig. 13.5. Schematics of a liquid nitrogen rotary tunnel.

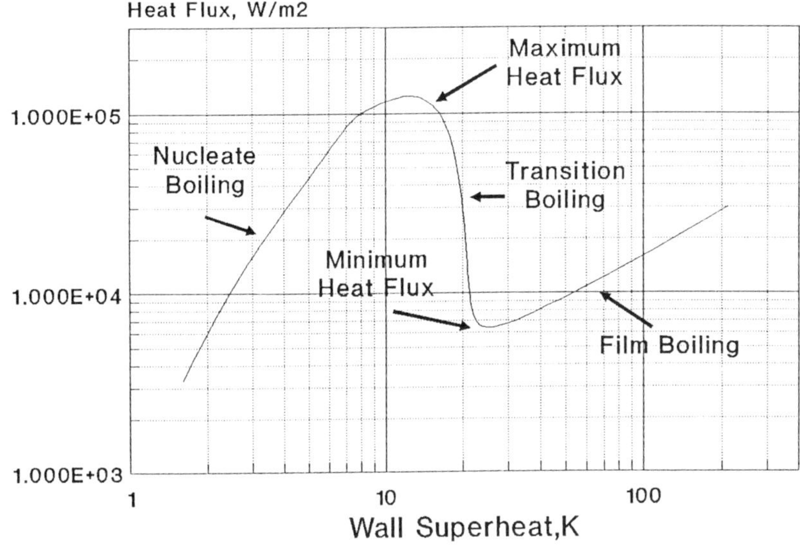

Fig. 13.6. Experimental pool boiling curves for liquid nitrogen.

The link between freezing rate and ice crystal size is well established, the faster the freezing rate, the smaller the ice crystals. For example, in cream freezing rapid freezing tends to lessen the amount of fat de-emulsified as the water in the cream freezes (Desrossier Tressler 1977). With raw yolk egg fast freezing and thawing result in less gelation than slow freezing and thawing (Stadelman Cotterill 1986) while freezing starter cultures in liquid nitrogen keeps the cultures viable for longer periods (Desrossier Tressler 1977).

To overcome the product deterioration much work has been done and is continuing with the freezing of liquids by immersion in, or direct contact with, liquid nitrogen. The liquid product being split up into drops to further enhance freezing rate. In the freezing and preparation of cultures this technique has become widely adopted and produces a free flowing granular product. Immersion equipment as described in the previous section of this paper has been used to freeze liquid egg and cream on an industrial scale.

However the following are a few limitations to the immersion technique.

Heat Transfer Rate Figure 13.6 depicts typical experimental pool boiling curves for nitrogen. With input product temperatures of typically 5°C and output temperatures of typically -40°C, temperature difference between the product and liquid nitrogen (-196°C) are in the range of 150 K-200 K, all within the film boiling regime of 20 to 28 kW per sq m of product surface area. The phenomena of film boiling limits the cooling rate in the process.

Low Bulk Density The rapid boiling of the liquid nitrogen exfoliates the liquid product which then freezes with nitrogen gas filled voids. The frozen product has an appearance not unlike granular insulant. Bulk densities can be as low as 400 kg per cubic m.

Bulk Thermal Conductivity The exfoliation reduces the effective thermal conductivity, thereby reducing the conduction of heat to the core of a bulk pack of frozen product. Consequently rapid thawing of the outer layer will occur during materials handling.

Liquid Nitrogen Consumption The extremely rapid freezing of the small droplets of liquid product limits the efficiency of the process as no advantage can be obtained from the use of the resultant cold nitrogen gas. Consumption of liquid nitrogen per unit of food product is therefore greater than with processes using pre or post cooling such as immersion freezing as described in the earlier sections of this paper. For example freezing water droplets to -40°C by direct immersion will require a theoretical minimum consumption of 2 kg of nitrogen per litre of water. Alternatively if the cold gas could be used to cool after immersion and in the process is warmed to -80°C, the theoretical minimum consumption would fall to 1.28 kg per litre of water.

The cooling rate obtained with liquid nitrogen is considerably less than that obtained using the cryo-biological technique of slamming. In this technique a cold plate of metal is brought into direct contact with the sample to be frozen and heat is conducted directly to the metal plate. Initial cooling rates are extremely high but reducing as a function of time.

A new development by Air Products PLC utilizes this technique in an indirect freezing method for liquid products which combats the above problems.

Indirect Liquid Freezing ("Cryostream")

Principles of Operation

A cut-away view of the cryostream freezer is shown in Figure 13.7. The process consists of a rotating stainless steel drum which is refrigerated on the inside surfaces by a recirculating turbulent flow of cold nitrogen gas, the gas flow is refrigerated by the addition of a liquid nitrogen spray, which vaporises to the cold nitrogen gas. By controlling the temperature of the recirculating gas stream the latent heat of the liquid nitrogen plus sensible heating of gas is obtained.

Liquid product is introduced to the surface of the drum as droplets (typically) and removed as frozen product after a portion of a single rotation. The thickness of the drum is calculated to provide a measure of cold storage such that the initial cooling rate is high as in the slamming technique. In addition the heat transfer to the recirculating gas stream is continuous. Once the product has been removed the drum is refrigerated again in the remaining part of a single revolution. When run as a continuous mode, any point on the stainless steel drum continuously follows the same temperature cycle.

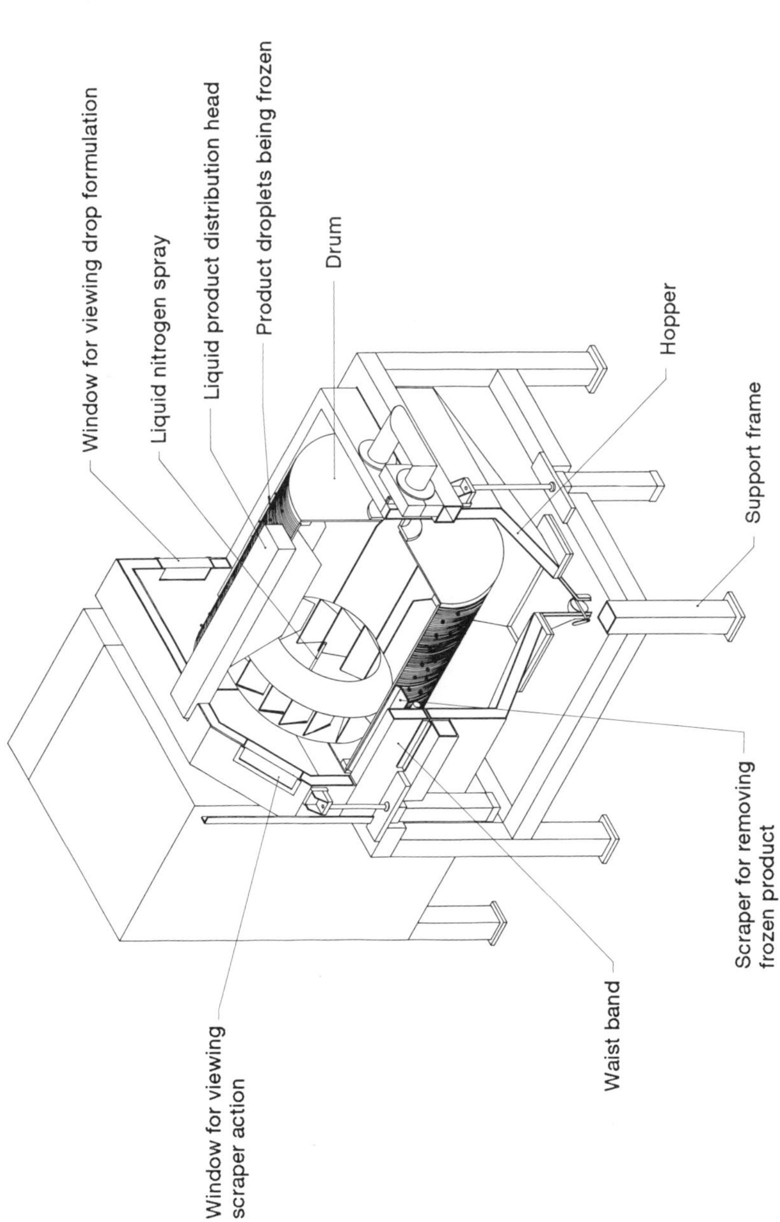

Fig. 13.7. Cryostream schematic.

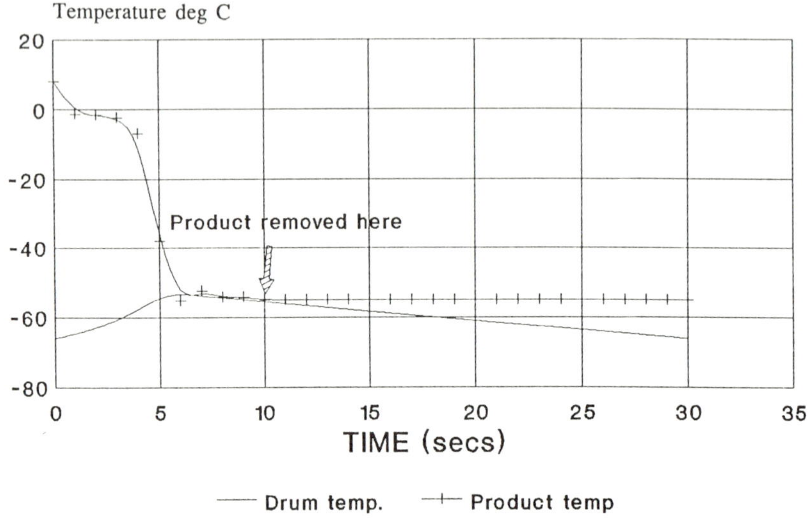

Internal Temp -120°C

Fig. 13.8. Cryostream high cooling rate.

Process Control

The drum rotation, the internal temperature, the portion of drum used for freezing and the product coverage are all parameters which are adjustable and could be used to vary the cooling rate and product temperature. In Figure 13.8 the internal drum temperature was -120°C and the product was removed after 120 degrees of revolution and was fully frozen in approximately 4 s. Figure 13.9 depicts a slower cooling rate with the same product. The internal drum temperature was -90°C and the product was removed after 270 degrees of revolution. In this case it is fully frozen in approximately 7 s.

Products

The frozen food product produced from this process is not exfoliated and exhibits bulk densities of between 600-700 Kg/m3. The problem of preferential thawing of bulk package surface layers is lessened by the improvement in bulk thermal conductivity.

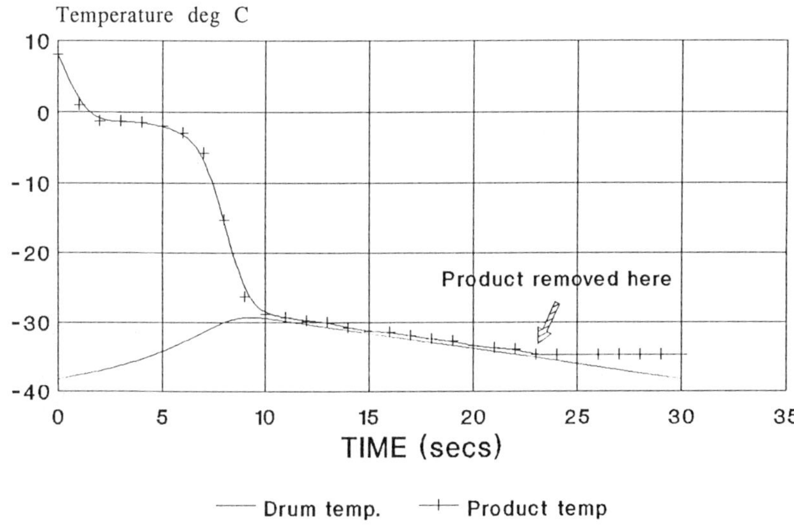

Internal Temp -90°C

Fig. 13.9. Cryostream Lower Cooling Rate.

Which keeps the outer layers below the melting temperature. Of particular application are products such as:

Liquid egg (whole, yolk and white);
Cream;
Speciality milks;
Starter cultures.

The process could also be a useful research tool for investigating the effects of different freezing rates.

Conclusion

Cryogenic techniques have found application for the industrial scale production of frozen solid and liquid foodstuffs. Processes are now available to meet the existing needs of the food industry. With the ever increasing consumption of frozen foods, driven by changing lifestyles and the emphasis on high quality additive free foodstuffs cryogenic technology will continue to offer solutions to the challenges of the frozen food industry.

Acknowledgements

To the author's employer, AIR PRODUCTS PLC, for permission to use various graphs, photographs and information.

References

Desrosier NW and Tressler DK (1977) Fundamentals of Food Freezing AVI Publishing Co,Westport, Connecticut pp 360-367, 384-386

Merte H and Clarke JA (1962) Boiling Heat-Transfer Data for Liquid Nitrogen at Standard and Near-Zero Gravity Adv Cryog Engng 7: 546-550

Stadelman WJ and Cotterill OJ 1986 Egg Science and Technology 3rd edition, Macmillan Publishers UK p 221

Chapter 14

Light Microscopy of Foodstuffs During Freezing and Thawing

M. R. McLellan, G. J. Morris, B. W. W. Grout, K. Hughes

Introduction

The post-thaw textural quality of many frozen foodstuffs is in some instances unacceptable, and in others sub-optimal. Poor textural quality is symptomised by a loss of firmness and resistance to bite, and often by fluid loss, the origin of which varies according to food tissue. In meat, drip loss is associated with breakdown in connective tissue (Love 1966) whilst in other foods, particulary fruit and vegetables, breakdown of cellular structure is the primary cause of quality decline (Reid 1987).

The firmness of plant-derived foodstuffs depends on cell turgor. Since cell turgidity requires a balance of osmotic pressure against hydrostatic pressure, pre-requisites for the maintenance of turgor are intact vacuolar and cell membranes to retain osmotic pressure, and cell walls to maintain hydrostatic pressure. As plant tissue ripens, senesces, or is subject to external stress, breakdown of cell and vacuolar membranes will occur, resulting in fluid loss from the cell membrane, and softening of tissue. The destructive effects of freezing stress on plant cell and endo-membranes have been well documented (Grout and Morris 1987). It is hardly surprising therefore, that acceptable post-thaw texture is often hard to achieve in frozen plant-derived food material where cellular organisation may well be sub-optimal prior to freezing due to ripening and post-harvest decay.

Nevertheless, studies in viable plant material have demonstrated that freezing damage can be minimised by altering freezing protocols, particularly by manipulating cooling rate. The responses of viable tissue to freezing at the cellular level are now known in some detail by cryobiologists as a result of a variety of microscopical, physiological and biochemical investigations. This knowledge has been put to practical use through the formulation of cryopreservation protocols for a range of cell types and in the

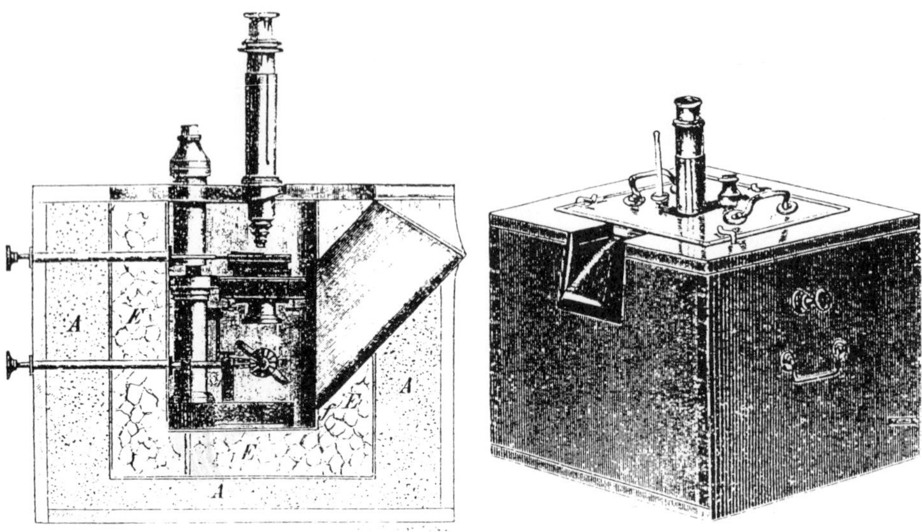

Fig. 14.1. The cryomicroscope system used by Molisch. An insulated box (A) surrounded an ice/salt freezing mixture (E), around an inner sleeve containing a microscope.

improvement of cold-hardiness in crop species. The study of damage to food tissue cells by freeze/thaw cycles is less well advanced, and more work is required in order that cellular responses are understood. Freezing protocols causing minimal cell and tissue damage must therefore be devised. The food industry has particularly neglected microscopical techniques in this regard. This neglect needs to be corrected, since microscopical techniques have been of demonstrable value in cryobiological studies. A range of microscopical techniques are available; freeze fixation and sectioning were used by Love (1966) and others to identify ice crystal size and location within food cells. Cryo-electron microscopy offers a means of examining ice crytal formation and damage at high magnification (see Wilson Chapter 8 this volume). The dynamic observation of cells during freezing (cryomicroscopy) is an additional method for studying the effects of freezing. This is emphasised here as a technique for the understanding and improvement of freeze/thaw protocols of food material. The use of cryomicroscopy in cryobiology is described, and its potential use in the food industry discussed.

Cryomicroscopy

Microscopical observation of chemical and biological systems during cooling (cryomicroscopy) was first reported by Molisch (1897 trans 1982). His "cryomicroscope" system comprised an insulated box containing an ice/salt freezing mixture, within which the microscope was placed (Fig. 14.1). Molisch used the apparatus to study the formation of ice crystals, and the location of ice in tissues during freezing.

Improvements in the design and construction of cryomicroscopes have paralleled the increasing sophistication of the light microscope. Cooling of specimens on a microscope stage can now be computer-controlled at exact cooling rates between 0.01°C and 100°C min. Fine control of the temperature of observed specimens, combined with the range of microscopical techniques now available, makes modern

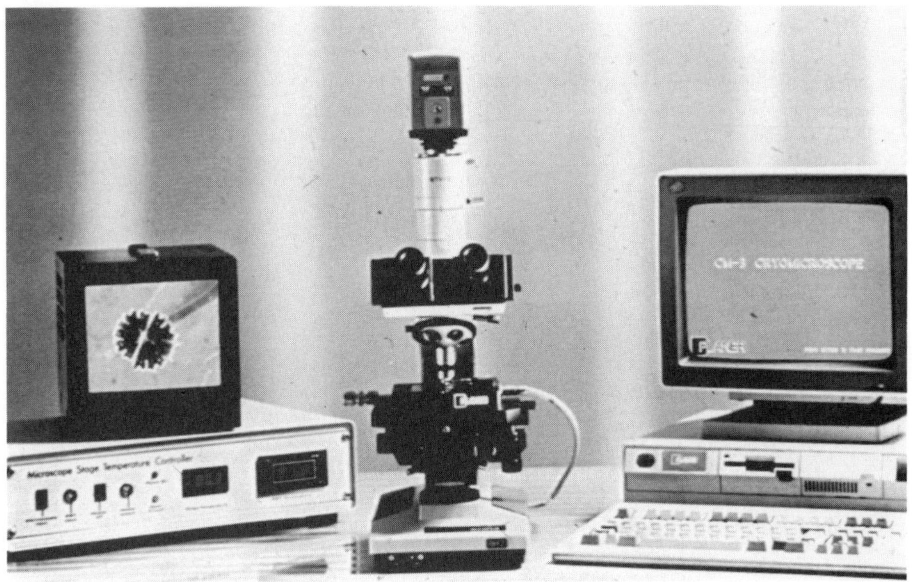

Fig. 14.2. The CM3 cryomicroscope. The software-driven cryostage/microscope system is interfaced to video and image analysis equipment. The system was designed jointly by Professor J.J. McGrath, Michigan State University, Cell Systems Ltd, and Planer Products Ltd.

cryomicroscopy an incisive tool in the study of cooling, freezing, thawing and heating on biological and chemical systems. Video microscopy allows information to be stored, and subsequently assessed using image analysis techniques. A typical modern cryomicroscope system is the Planer CM3 (Fig. 14.2).

The basic elements involved in controlling the temperature of a sample on the cryomicroscope can be divided into several subsystems (Fig. 14.3).

The desired sample temperature is created by the reference signal generator, which is compared with the actual sample temperature measured by the temperature sensor. A controller unit produces an output which is proportional to the difference between the desired and actual temperatures, in the form of an error signal to a power amplifier. The amplifier outputs electrical power to the heater stage (Fig. 14.4).

The heater stage is coated with a thin metal film, which causes supplied electrical energy to be converted to thermal energy. The temperature sensor directly below the sample detects the new sample temperature, sending this information to a temperature conditioner, creating a closed feedback loop (Fig. 14.3). For cooling, a heat flux is created by cooling a metal plate contacting the heater stage using a refrigerant flow (Fig. 14.4). A fuller description of the theory and practice of cryomicroscopy can be found in McGrath (1987).

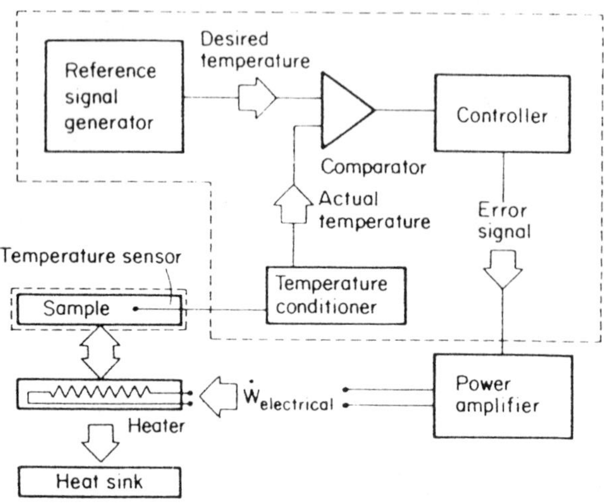

Fig. 14.3. The basic elements of a temperature controller for a modern cryomicroscope.

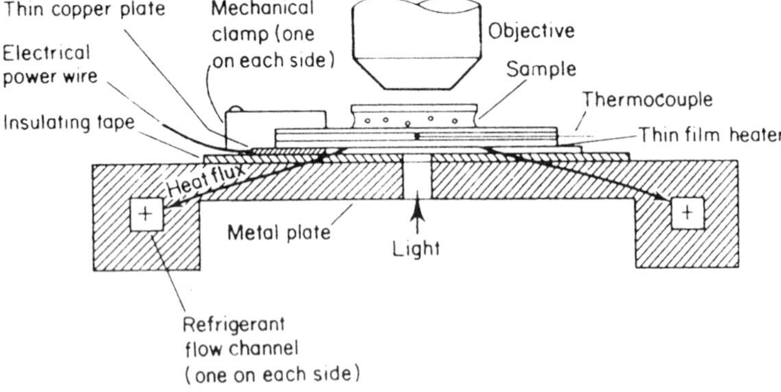

Fig. 14.4. Heat transfer stage of a modern cryomicroscope.

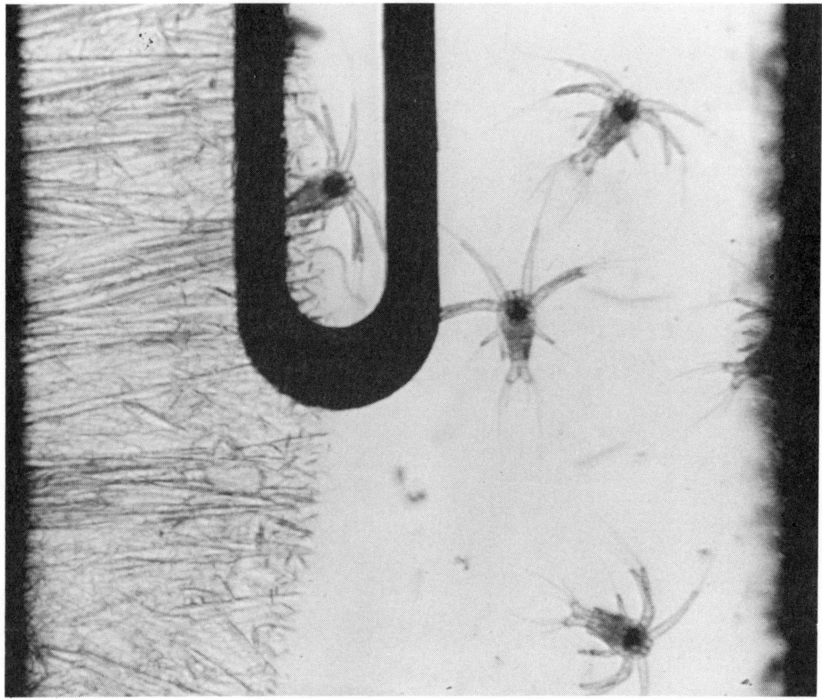

Fig. 14.5. Cryomicrograph of shrimp larvae, demonstrating an ice front at the thermocouple junction. The freezing point of the bathing solution was -2.3°C.

Potential Uses of Cryomicroscopy in the Food Industry

Cryomicroscopy can be used to determine a number of events occurring as a result of lowering temperature, including:

a Temperature of ice nucleation in a tissue or bathing solution;
b Size, shape and location of ice crystals around and within cells or tissues;
c Rate and mechanism of propagation of ice crystals;
d Mechanical effects of extracellular ice crystals on cells/tissues;
e The volumetric response of cells during freezing;
f The morphological alterations occurring during freeze/thaw cycles;
g The effects of cooling rate, chemical additives, and other factors on a - f above.

The usefulness of investigating each of these factors for food freezing is discussed, with reference to their previous implementation either in cryobiology or food freezing.

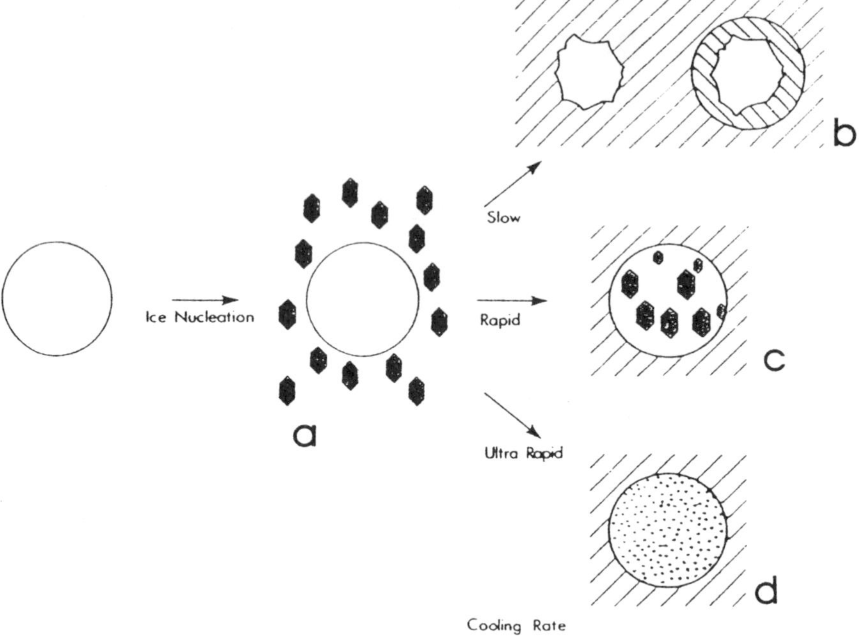

Fig. 14.6. Schematic diagram of the ice nucleation patterns in cells at different cooling rates. See text for explanation.

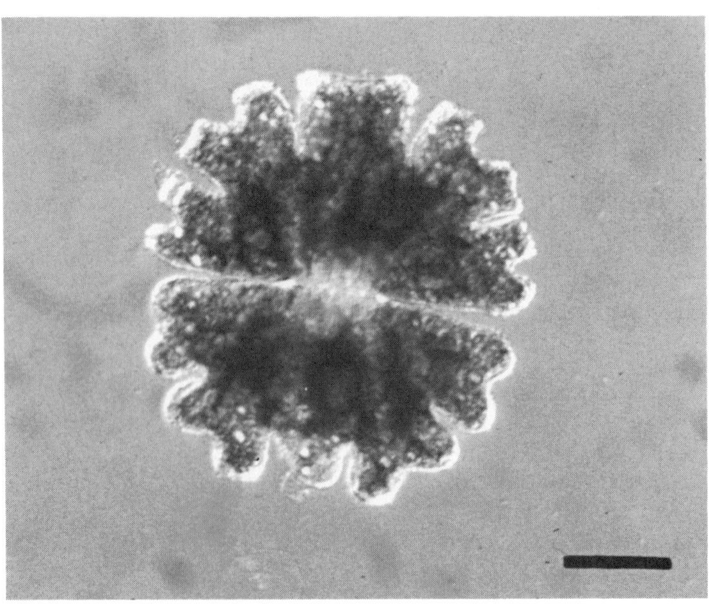

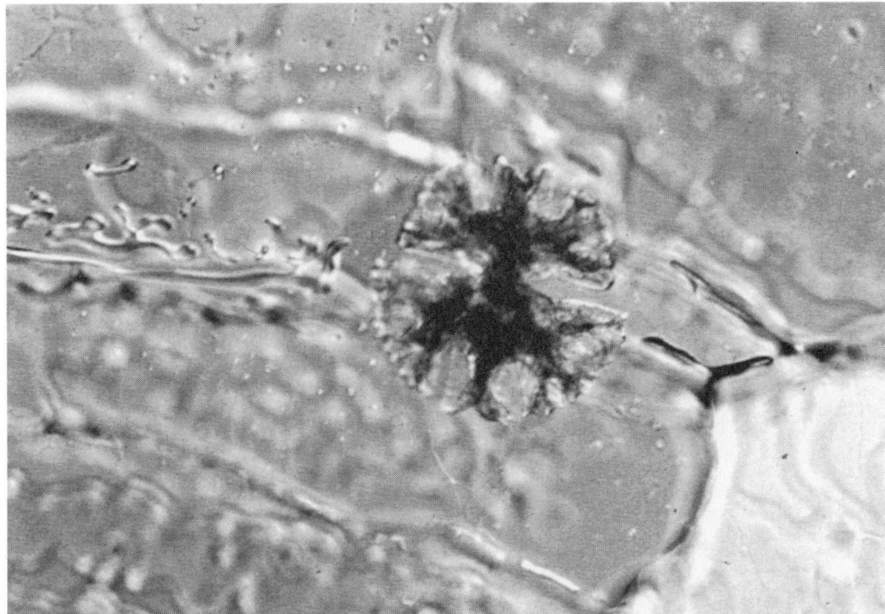

b

c

Fig. 14.7. Cryomicroscopy of *Micrasterias rotata.* **a**) control; **b**) -10°C, frozen at 2°C min^{-1}; **c**) -10°C, frozen at 30°C min^{-1}. Scale bar = 20 μm.

Temperature of Ice Nucleation

The thermocouples used in the construction of a cryostage are accurate to ± 0.1°C, and offer the potential of determining the freezing point of solution, cell, or organism placed immediately above them. In a series of experiments on shrimp larvae, cryomicroscopy was used to determine the freezing point of external solutions and that of the larval body fluid (Fig. 14.5).

The technique is a rapid means of measuring the freezing point of solutions and emulsions used in the food industry, or comparing the freezing points of heterogenous cells in food tissue samples.

Location of Ice in Tissues during Freezing

Ice nucleation can be located within (intracellular ice) or outside cells (extracellular ice). In cell populations, freezing is usually initiated in the bathing medium surrounding the cells (Fig. 14.6a).

If ice remains extracellular, the bathing environment becomes hypertonic as a result of an increase in solute concentration. Providing the cells are viable, and the surrounding cell membrane is semi-permeable, water will be lost from the cell by osmosis, and cell shrinkage will occur (Fig. 14.6b). Alternatively, extracellular ice will trigger intracellular ice formation (IIF) causing the organelles and cytoplasm of the cell to freeze (Fig. 14.6c). The probability of IIF occurring is usually a function of cooling rate; as cooling rate is increased, less time is available for water to leave the cell osmotically. Supercooling of intracellular water will lead to an increased probability of IIF. As cooling rate increases beyond the threshold cooling rate required for IIF, the size of intracellular ice crystals diminishes (Fig. 14.6d). Manipulation of the cooling rate for cryopreservation protocols is aimed to avoid IIF, since in all but rare cases, viability is lost due to organelle rupture and loss of cellular compartmentation. Equally, excessively slow cooling rates are avoided, since osmotic damage can result from prolonged exposure to extracellular hypertonicity. An example of the effect of cooling rate on cellular viability is given in Figure 14.7.

Micrasterias rotata, a freshwater algal species was frozen at three different cooling rates on a cryomicroscope stage, allowing the location and size of ice crystals to be visualised. Cooling at 2°C min^{-1} caused cell shrinkage and hypertonic damage subsequently causing cell death (Fig. 14.7b); cooling at 30°C min^{-1} caused intracellular ice to form (note cell darkening) which caused irreversible damage to the cell (Fig. 14.7c). Cooling *Micrasterias* cells at 10°C min^{-1} caused no IIF, and minimised hypertonic exposure, resulting in minimal cell damage, and a high percentage of the cell population surviving freezing (Morris, Coulson and Engels, 1986).

In cells comprising certain food tissues, similar responses during freezing to those obtained in cell populations in suspension are obtained. During slow freezing of tomato sections, Luyet (1968) observed extracellular ice formation at slow cooling rates leading to hypertonic shrinkage of cells. However, when the cooling rate was increased, intracellular ice resulted, causing tearing of cytoskeletal elements, and other injurious effects. Similar events have been observed in fish muscle by Love (1966) using freeze-fixation techniques. However, in many food tissues, freezing will be exclusively intracellular, independent of cooling rate employed. This is because the bulk of tissue water is intracellular, with little or no extracellular fluid to drive hypertonic removal of water from cells. This is true, for example, of many fruits and hypertonic removal of water from cells. This is true, for example, of many fruits and vegetables. In addition, hypertonically-driven exosmosis occurs only in cells where the cell membrane is semi-permeable; in ripe or harvested plant material, breakdown of cell membrane structure/function is likely. The contribution of cellular organisation to the freezing behaviour of food tissue is considered later in the chapter.

Nucleation of Ice Crystals

Ice nucleation at or near the freezing point of a solution will not occur unless an intrinsic nucleating particle or extrinsic agent (e.g. mechanical seeding) of nucleation is present. Where nucleators are absent, supercooling of solutions below the freezing point occurs resulting in "cold shock" damage to some cell types (McLellan *et al.* 1984). The extent of undercooling in food bodies has not been well researched, therefore it is not well established whether undercooling causes detrimental effects in frozen/thawed foods. Certain food objects, including some berries demonstrate undercooling of the whole mass. However most food tissues (e.g. grapes, strawberries) harbour efficient nucleators of ice often in the epidermis of the tissue; these may be bacteria or macromolecules. Whilst these particles prevent undercooling of the entire fruit, undercooling of individual parenchyma cells probably does occur. The effects of reducing the probability of undercooling in individual cells in food bodies on the textural quality of frozen/thawed material have yet to be investigated.

Propagation of Ice Crystals

Ice nucleation occurs generally at the surface of the food tissue, with ice fronts progressing towards the centre of a food object. Though an obvious effect, this mechanism of ice propagation has a number of implications with regard to the rate at which food is frozen. If a very rapid cooling rate is employed, the cell layers of the tissue interior will remain warm relative to the peripheral cell layers. Heat gradients across a tissue can lead to physical shearing, and loss of textural quality and appearance (several authors, this volume).

At the cellular level, the spread of ice within a food tissue will depend on its cellular organisation. Many foodstuffs contain cells which are senescent (e.g. post-harvest fruit); dying (e.g. post-harvest shoots); decaying (post-slaughter meat or fish) or killed (e.g. blanched vegetables). In such tissue, intracellular compartmentation and membrane semi-permeability will be in varied states of breakdown. This kind of tissue is exemplified by a heat-treated onion epidermis (Fig. 14.8a). Ice spreads through the tissue uninterrupted via a single ice front (Fig. 14.8b, c), resulting in large ice masses throughout the tissue.

Similar observations were made by Brown and Reuter (1974) on freeze-damaged cucumber sections. The size of ice crystals in such tissue will be dependent on the numbers of nucleation sites; the greater the number of nucleation sites, the smaller will be the crystals. To some extent, this can be controlled by cooling rate; in emulsions such as cream, long needle shaped crystals grow from the sample periphery to the interior at slow cooling rates; smaller more numerous crystals occur at faster cooling rates, as a result of an increased number of nucleation sites.

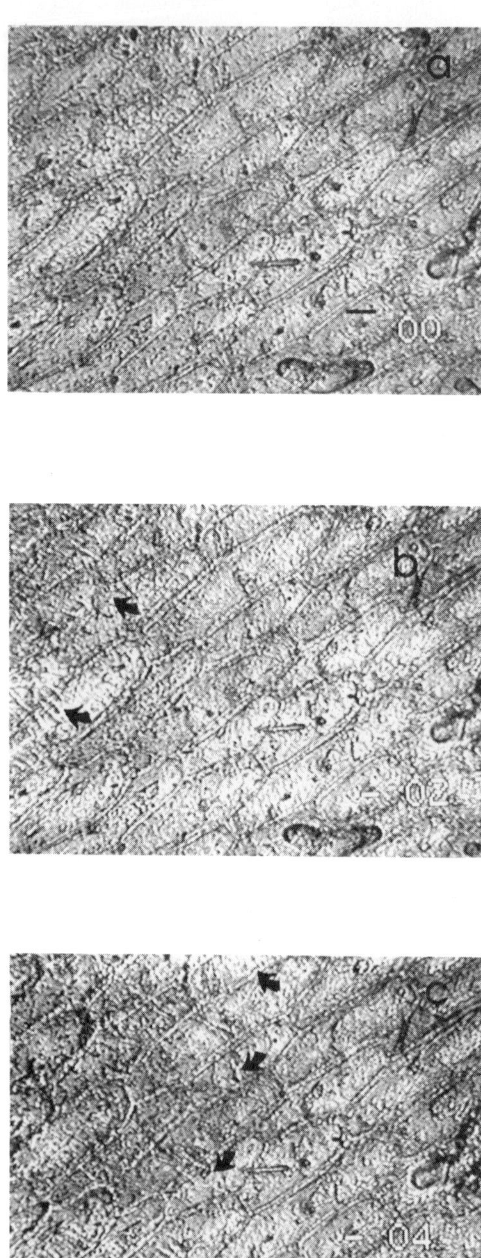

Fig. 14.8. Cryomicroscopy of heat treated red onion epidermis cooled at 1°C min^{-1}; **a**) 0°C; no ice present; **b**) -2°C, extracellular ice front (arrows) **c**) -4°C progression of extracellular ice front(arrows). Scale bar = 20 μm.

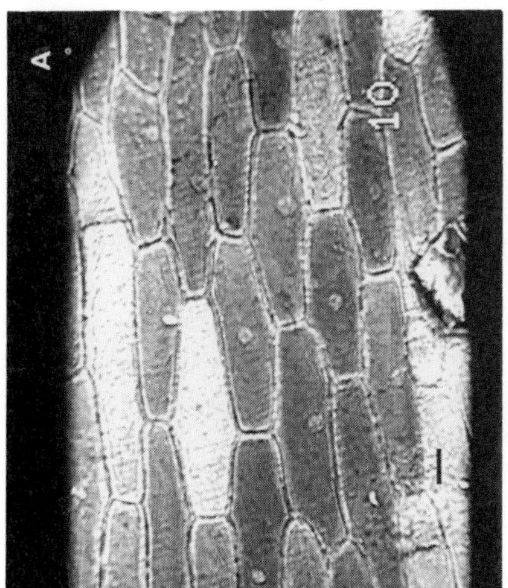

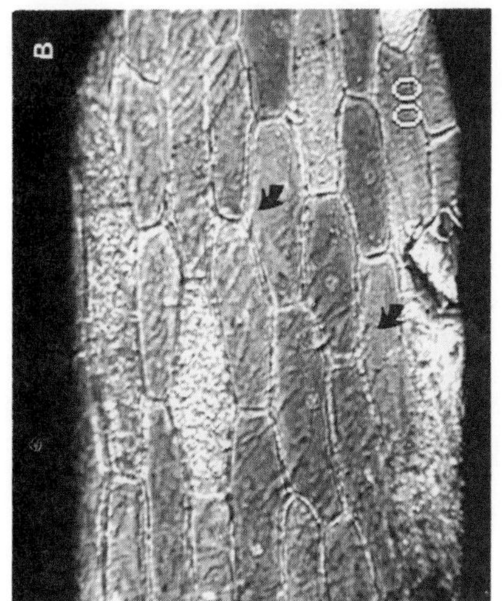

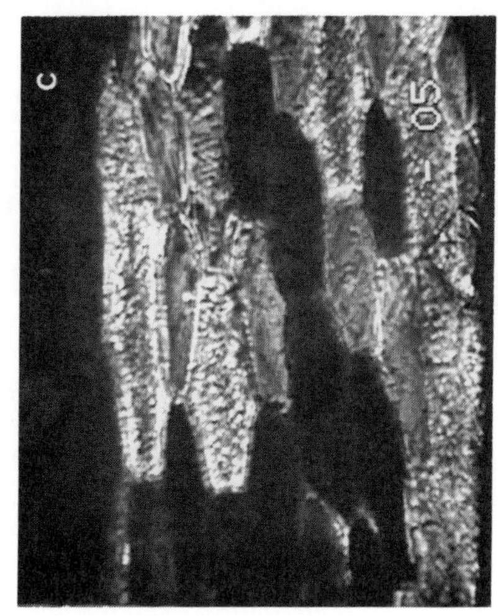

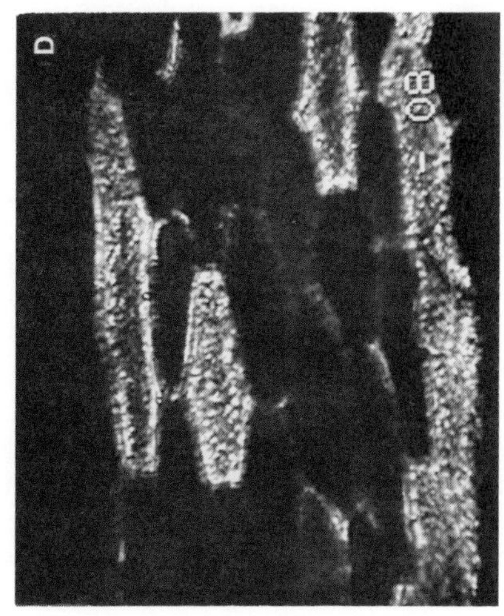

Fig. 14.9. Cryomicroscopy of fresh red onion epidermis, cooled at 1°C min^{-1}; **a)** 10°C; **b)** 0°C, progression of extracellular ice front (arrows); **c)** -5°C; darkening of cells indicates intracellular ice; **d)** progression of intracellular ice through tissue. Scale bar = 30 µm.

In tissue where cellular organisation is intact, (for instance in unripe fruit or fresh vegetables) the spread of ice is more complex. In fresh cucumber slices, Brown and Reuter (1974) observed that the mechanism of nucleation was by internal freezing of cells which seeded intracellular ice in neighbouring cells, and subsequent propagation of ice throughout the tissue proceeded in this manner. Fresh red onion tissue contains sufficient extracellular water to promote ice formation external to cell membranes and in intercellular spaces (Fig. 14.9a, b).

During further cooling, individual cells freeze internally, which then act to seed intracellular ice in neighbouring cells (Fig. 14.9c, d). In viable tissue, intracellular ice is invariably associated with cell death, there being few examples of cell types able to recover from this state. Exceptions include certain algal (Morris 1981) and fungal (Morris *ert al.*1988) species. Viability is of little concern to freezers of food material; however, the effects of intracellular ice on cell structure is of critical importance to the post-thaw response of cell-based material. Therefore the damaging effects of intracellular ice in food tissue should be examined, in order that those effects be ameliorated in the design of freezing protocols.

Size and Intracellular Location of Ice Crystals

Control of ice crystal size within frozen food tissue has been identified by the food-freezing industry as critical to post-thaw textural quality. Evidence suggests that relatively slow freezing causes large intracellular ice crystals, and relatively fast freezing causes small intracellular ice crystals (e.g. Love 1966), an effect also general to dispersed viable cell populations (Fig. 14.6). Small ice crystals are deemed to be minimally damaging to cells, by avoiding the rupture of both cell membranes and internal organelles. This has led to the conclusion that the fastest possible rates of cooling which avoid gradient effects and cracking must lead to the optimal textural preservation of food post-thaw. Studies comparing the size and subcellular location of intracellular ice in food cells with post-thaw textural characteristics are lacking. Hence the optimal size/subcellular location of ice crystals for cellular and therefore tissue preservation cannot be estimated or controlled.

Cryobiological studies have indicated that different regions of the cell can contain different sizes of ice crystal depending on cooling rate. Grout, Willison and Cocking (1973) demonstrated that vacuolar ice crystals are larger than cytoplasmic ice crystals in rapidly frozen tomato protoplasts. Studies comparing the size and sub-cellular location of ice crystals with post-thaw texture in food tissue are clearly required and how these factors are affected by cooling rate. It is possible that the potential benefits of causing minimum ice crystal size within a cell have to be offset against other potentially damaging factors. This should include a degree of undercooling and length of exposure to high sub-zero temperatures where ice-growth and recrystallisation is likely. This possibility is discussed further by Grout, McLellan and Morris (Chapter 9 this volume).

In order to test this hypothesis, microscopical analysis of food tissue must be made. Intracellular ice crystal size can be estimated using cryomicroscopy; Luyet (1968) was able to view small intracellular ice crystals following plunging tomato tissue to 120°C. Other methods for studying intracellular ice in detail are cryo-scanning electron microscopy (cryo-SEM, see Wilson Chapter 8 this volume), or by freeze-fixation and sectioning (e.g. Love 1966).

If a protocol could be devised whereby intracellular ice crystal formation minimised rupture of subcellular compartments, improved cellular preservation would occur. This would remove the need to blanch certain vegetables prior to freezing. At present, blanching is done to thermally inactivate enzymes which when mixed with substrates cause deteriorative changes in cell and tissue. Since conventional freezing causes breakdown of endomembranes which usually separate such enzymes and substrates, a failure to blanch would result in degraded products appearing in frozen/thawed food. A freezing protocol which totally or partially prevented endomembrane breakdown would prevent undesirable post-thaw mixing of enzyme/substrates, thereby rendering the blanching process unnecessary.

Recrystallisation

It has been well established that ice crystals grow or recrystallise at high sub-zero temperatures (Franks 1985), and that such processes can be observed dynamically using cryomicroscopy (e.g. Luyet 1968). This phenomenon has several important implications for the effective preservation of frozen food, both during the freezing process and subsequent storage. During much conventional food freezing, latent heat of fusion is dissipated slowly, resulting in a slow cooling profile, with a long period of time where frozen food is exposed to sub-zero temperatures where ice recrystallisation is likely. Processes which reduce exposure to such temperatures, whilst avoiding the problems associated with excessive cooling are likely to be optimal for textural preservation. The implications of this conclusion for the bulk freezing of food are considered in Chapter 9 in this volume (Grout, McLellan and Morris).

Morphological Alteration of Cells/Emulsions

Cryomicroscopy offers the possibility of following events occurring during a complete freeze/thaw cycle in a foodstuff sample. In addition to visualising the formation of ice, the morphological appearance of cells, solutions and emulsions can be monitored, and damage symptoms noted. Thus, freezing protocols can be assessed at the cellular level, enabling information concerning the potentially damaging or ameliorative properties of each protocol to be gained rapidly. In addition to cell damage, rupture or leakage, emulsion breakdown during thawing can be assessed. The extent of emulsion breakdown can be assayed by monitoring the number and size of aqueous or lipid droplets appearing in a fixed sample volume. For example, in a study of the effect of thawing rate on frozen single cream, the emulsion of cream frozen at 1°C min^{-1} and thawed rapidly at 40°C min^{-1} began to break down. At 18°C during rewarming (Fig. 14.10b) droplets formed with destability maximised at 25°C (Fig. 14.10c). Reduction of the thawing rate at any specific cooling rate is directly related to preservation of emulsion stability (Morris and McLellan, unpublished data).

Cryomicroscopy is also suitable for assessing the effects of storage time under optimal or sub-optimal storage conditions. Any detrimental changes during storage can be identified and monitored at the cellular/microscope level by transferring frozen material to a cryostage pre-set at any required storage temperature.

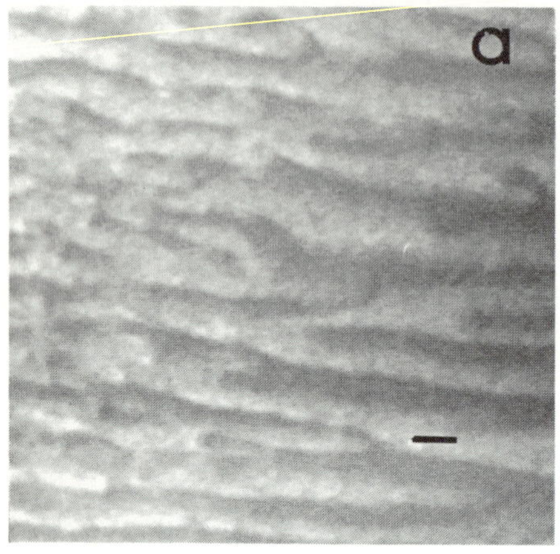

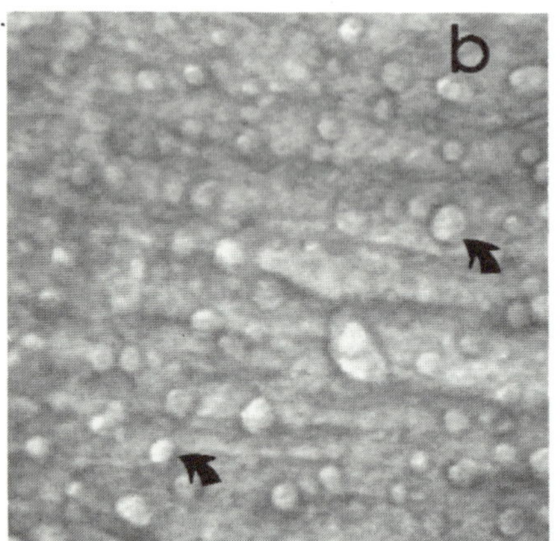

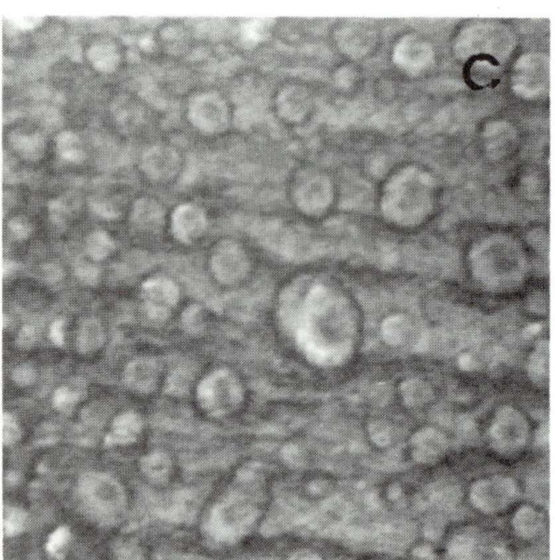

Fig. 14.10. Cryomicroscopy of single cream. **a)** frozen at -10°C, following cooling at 1°C min $^{-1}$;**b)** thawing at 40°C min^{-1} ,18°C, emulsion breakdown and appearance of droplets (arrows); **c)** 25°C following thawing, growth of droplets. Scale bar = 30 μm.

Response of Foodstuffs to Pre-freeze Processing

Many foodstuffs are heat-killed, blanched, or chilled prior to freezing, processes which clearly affect the state of cellular preservation. Difficulties with post freeze/thaw texture or other vital characteristics of food can often be ascribed to pre-freeze temperature excursion. The indentification of the root of textural problems at the cellular level can be studied using cryomicroscopy, by direct observation of cells through pre-freeze treatments, followed by freeze/thaw cycles. Sites of damage and likely ameliorative measures can be quickly identified, and subsequently tested in bulk freezer experiments.

Combination of Microscopical Methods

In many instances, a combination of microscopical methods will yield most information regarding the behaviour of foodstuffs at the cellular level. All microscopical methods have inherent advantages and drawbacks (see Wilson Chapter 8 this volume), a combination of methods is therefore least likely to produce artifactural results. One effective way of combining light and electron microscopy in the study of frozen food uses both cryomicroscopy and cryo-scanning microscopy as established by Roberts *et al.* (1987). The technique enables the direct observation of cells and tissues during freezing, and the same material is preserved and mounted for electron microscopy. Thus, cellular and ultrastructural information during freezing can be combined, enabling a detailed analysis of the freezing protocol under study.

Acknowledgement

The authors thank Anita Boyce for typing the manuscript.

References

Brown MS and Reuter FW (1974) Freezing of nonwoody plant tissue III. Cryobiology 11:185-194

Franks F (1985) The technology of metastable water. In: Franks F Biophysics and Biochemistry at Low Temperatures. Cambridge University Press pp 167-187

Grout BWW, Willison JHM, Cocking EC (1973) Interactions at the surface of plant cell protoplasts; an electrophoretic and freeze-etch study. Bioenergetics (1973) 4:311-328

Grout BWW and Morris GJ (1987) Freezing and cellular organisation. In: Grout BWW, Morris GJ (eds) The Effects of Low Temperatures on Biological Systems. Edward Arnold pp 147-174

Love RM (1966) The freezing of animal tissue. In: Meryman HJ (ed) Cryobiology Academic Press New York pp 317-406

Luyet BJ (1968) Investigations on the freezing and freeze-drying of selected fruits and vegetables. Technical Report 69-8-FL. US Army Natick Laboratories, Natick, Massachusetts

McGrath JJ (1987) Temperature-controlled cryogenic light microscopy - an introduction to cryomicroscopy. In: Grout BWW, Morris GJ (eds) The Effects of Low Temperatures on Biological Systems. Edward Arnold pp 234-267

McLellan MR, Morris GJ, Coulson GR, James ER, Kalinina LV (1984) Role of cytoplasmic proteins in cold shock injury in *Amoeba*. Cryobiology 21: 44-59

Molisch H (1982) Investigation into the freezing death of plants. Cryoletters 3: 331-390

Morris GJ (1981) Cryopreservation. Institute of Terrestrial Ecology 1-27

Morris GJ, Coulson GE and Engels M (1986) A cryomicroscopic study of *Cylindrocystis brebissonii* de Bary and two species of *Micrasterias* Ralfs (Conjugatophyceae, Chlorophyta) during freezing and thawing. J exp Bot 37 (179): 842-856

Morris GJ, Smith D, Coulson GE (1988) A comparative study of the changes in morphology of hyphae during freezing and viability upon thawing for twenty species of fungi. J gen Microbiol 134: 2897-2906

Reid DS (1987) The freezing of food tissues. In: Grout BWW, Morris GJ (eds) The Effects of Low Temperatures on Biological Systems. Edward Arnold, pp 487-488

Roberts S, Grout BWW, Morris GJ (1987) Consecutive observation of a frozen cell sample by cryogenic light microscopy and cryogenic scanning electron microscopy. Cryoletters 8: 122-129

Steponkus PL, Evans RY and Singh J (1982) Cryomicroscopy of isolated rye mesophyll cells. Cryoletters 3: 101-114

Chapter 15

Continuous Monitoring of Cryogen Consumption During Freezing of Foodstuffs

E. M. A. Willhoft

Introduction

At the turn of the century liquid air was of academic curiosity. In 1878 a paper appeared in Nature describing a method of producing liquid air by the hydraulic compression to 1000 bars of air according to L Cailletet and RP Pictet. The crude equipment used by them is shown in Figure 15.1. The compression and sudden expansion resulted in droplets of liquid air being formed.

In the thirties the refrigerant properties became exploited commercially and an early road tanker is shown in Figure 15.2. In those days - and even in the mid sixties - the cryogen was in the form of liquid air rather than the separated nitrogen. I recall in the sixties as a post-graduate student having to tell the difference between liquid air and liquid nitrogen by the colour. Oxygen in the liquefied air conferred a blue colour and if it splashed any part of the unprotected body, which was a constant hazard during transfer to a Dewar flask, it immediately produced burn marks, whereas (liquid nitrogen) LN_2 simply evaporated off the skin leaving it unharmed. For reasons of self preservation you wouldn't dare immerse your hand in liquid air or liquid oxygen; but a very brief immersion in LN_2 did no harm.

LN_2 has a slightly lower boiling point (-196°C) than LO (-183°C), and it is for this reason that formation of puddles of LN_2 from a cryogenic operation should be guarded against, otherwise oxygen from the atmosphere condenses which can produce a dangerous hazard when in contact with a readily oxidisable material such as food.

LN_2 is now transported in large vacuum insulated tankers and stored in vacuum-insulated tanks at pressures between three and five atmospheres. The higher the storage pressure, the more cold is lost in flash heat in coming down to atmospheric operating pressure. A typical cryogenic food freezing system is shown in Figure 15.3.

The purpose of this paper is to highlight ways of reducing cryogen operating costs, principally by monitoring its consumption whilst freezing. For various reasons the

Fig. 15.1. Production of liquid air by L Cailletet and RP Pictet in 1877. Air was compressed hydraulically to 1000 bar and then allowed to expand rapidly which resulted in partial liquefaction. (Source: Messer Griesheim GMBH)

Fig. 15.2. Road tanker used for delivering liquid air during the 1930's. (Source: Messer Griesheim GMBH)

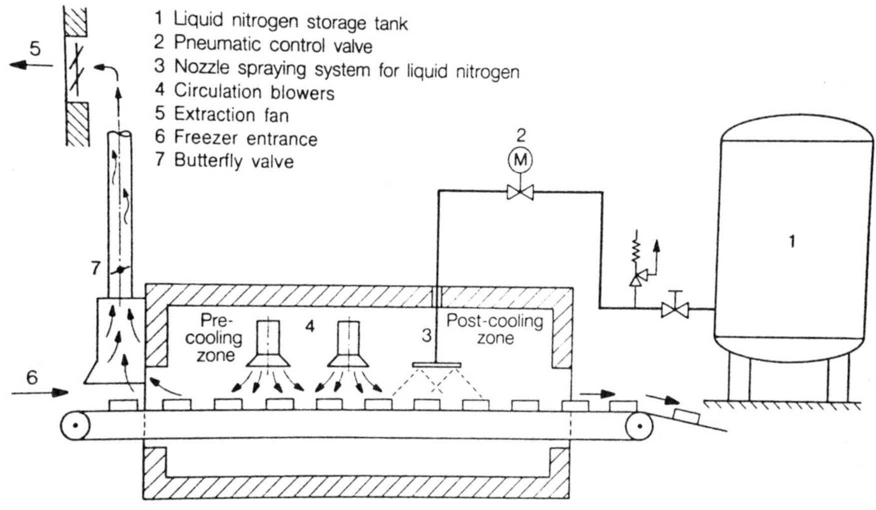

Fig. 15.3. Line diagram of LN_2 freezing system. (Source: Messer Griesheim GMBH)

technology of monitoring the rate of consumption at the time of freezing has not been encouraged by the gas companies, let alone been made available to the end user. The present paper identifies a simple approach to monitoring on-line the rate of gas consumption during a cryogenic operation, and is not confined to LN_2. The technology is equally applicable to CO_2 freezing for reasons that will be discussed later.

The Cost of Freezing Cryogenically

There are valid and invalid reasons for using cryogens in food freezing, when viewing an application from the point of view of the food processor. The subject is a wide one and cannot be fully covered in this chapter. Payment for an expendable processiong aid, including water, electricity, fuel gas etc, is assumed through custom and practice to be based on metering by the end user. Without monitoring, the performance of plant with regard to efficiency obviously cannot be assessed. However, with nitrogen freezing there is no monitoring facility available other than by use of a weighbridge for the tanker, or load cell placed under the storage tank. Filling a tank can take 2 h or 3 h (frequently with leakage), and in the meantime, the chances are that LN_2 is being removed for freezing products which then leaves some doubt as to the amount of gas transferred from the tanker.

Most tankers now have a flow-metering device connected between them and the storage tank so that a measure of the amount of LN_2 transferred is indicated by the in-line meter. It is claimed by the manufacturers of these meters - generally of the vortex-shedding type - that they are accurate to within 1%. What does that mean? It generally means that such an accuracy is established under idealised conditions using a liquid such as water or a fluid in the gaseous state - i.e. a monophase fluid. LN_2 is a fluid co-existing in the liquid and gaseous phases under the conditions of transfer from the tanker (at high pressure) to the lower-pressure storage tank. The vortex shedding meter relies for its effect on a monophase fluid - i.e. either liquid or gas, but not a mixture. Measurement of LN_2 content from a differential pressure within the tank assumes, incorrectly, a constant gas pressure set to correct the hydrostatic pressure of the varying height of liquid within the tank. Claims on the accuracy of measuring the amount of LN_2 transferred from the tanker are, thus optimistic.

The approaches adopted in practice have some value in giving an indication of when to re-charge a tank to maintain continuity of manufacture. What they do not do is monitor the rate at which LN_2 is consumed during freezing. Even with load cells placed under the storage tank, because of the large gross mass involved relative to the liquid removed during short time scales, the rate of removal of cryogen pales into insignificance. The analogy is almost on a scale of trying to measure the weight of the captain of a ship by measuring the displacement with him aboard and with him off the ship; theoretically feasible, but practically very difficult.

The quarterly cost of LN_2 for an end user can exceed £25K. Payment of these huge amounts must surely justify a facility for continuous measurement of the LN_2 ratio (LNR) to provide a means of operating the tunnel cost-effectively. The LN_2 costs imposed by the gas company for an agreed unit charge can be broken down as follows:

- i The amount transferred to the storage tank;
- ii The amount evaporated through boiling during road transport, which can take several hours;
- iii The spillage in transfer from the tanker to the storage tank (which can take up to 3 h);
- iv The amount evaporated during storage in the nitrogen tank. (A good tank loses 0.1% of the LN_2 every 24 h, or more than a third of a full tank each year through boiling losses.)

If the amount delivered is determined by the difference in weight of the tanker immediately before and after transfer at the customer's site, this will yield a different result than if the weight difference is obtained at the gas supplier's premises; unless the driver corrects for the fuel weight he had before starting on a journey as well as for evaporative losses incurred during the round journey.

The ideal situation from the customer's point of view is to install a facility that measures LN_2 whenever the tunnel draws on it. In other words, to ensure that the major cost variable is precisely monitored and controlled. It is a matter for the customer to determine who pays for transfer and delivery losses. Here we describe the computerised monitoring of cryogen whenever one or more tunnels is in use during freezing. It is referred to as the Computerised Monitoring System (CMS).

A small number of tunnels have a facility for controlling the extract-duct fan speed by linking the r.p.m's to the LN_2 demand. In other words, to use the demand signal for the gas valve to operate the extract fan speed. For a variety of reasons the relationship between fan speed and extract rate of cryogen at the exit of the tunnel is complex, and at certain speeds, control becomes ineffectual, particularly if air dilution is downstream of the entrance to the extract duct. If the speed of the extract fan exceeds a certain value then cavitation occurs. The situation is further aggravated by the vastly different temperatures - and, hence densities - of the spent cryogen gas and the entrained atmospheric gases at or close to the entrance of the extract ducting. Reducing the rpm of the extract fan in proportion to a reduction in cryogen demand will not have the same effect on the extraction of spent cryogen unless controlled by some other parameter.

The objective of reducing the extraction rate in total harmony with the thermal load is achievable, in principle, by analytically monitoring the cryogenic gas within the extract ducting and controlling the fan speed from the gas composition, but not allowing it to fall below a pre-set extraction rate - this then allows for idiosyncratic behaviour in the relationship between fan speed and thermal load within the tunnel and reduce waste in cryogen useage due to excessive gas extraction. Monitoring gas composition in the extract duct is, in fact a key step in quantifying cryogen consumption on a continuous basis when linked with a continuous measurement of total flow rate and temperature.

Monitoring the cryogen useage rate is the essential precurser for setting up control loops to eliminate inefficient tunnel performance. Loops may even be linked with a process operation up-stream or down stream of the tunnel, reflecting the possibility that gross inefficiency may be linked with a process step outside of the control of the freezing tunnel. This point needs to be made to counter claims made by gas companies that their tunnels do not operate inefficently. The high LN_2-waste levels sometimes found, according to the gas companies, are either caused by the operator, or plant unconnected with the freezing tunnel. Both these allegations are irrelevant to the justification for installing a monitoring facility. If excessive gas consumption, for whatever reason, can be reduced and, furthermore, provide a check and possible challenge to quarterly charges imposed by the gas company, then the add-on technology becomes justified. It becomes very cost-effective if the nett reduction in gas charges greatly exceeds the cost of the monitoring equipment.

Principal of Operation of the CMS

For reasons already covered it is virtually impossible economically, to measure accurately a varying rate of flow of cryogen downstream of the storage tank, based on measurements on the boiling LN_2. Once the cold content of the cryogen has been transferred to the food product, the spent cryogen leaves the tunnel as a gas. Indeed all of the gas must be extracted to waste quantitatively, otherwise problems of health and safety can arise. It is this simple fact that provides a practical means of using conventional gas-flow metering technology, well established in other fields, for measuring accurately the consumption of cryogen when combined with use of an oxygen analyser and a temperature-sensor.

At the position of extraction, the spent cryogen is at about 45°C below the temperature of the pre-frozen product. Unless the spent gas is deliberately diluted with atmospheric air, the extract fan motor will seize up. Consequently, the spent gas should be diluted with between three and four volumes of air for every volume of cryogen drawn to waste. In other words, the gases extracted to waste will consist of cryogen (nitrogen or CO_2) and atmospheric gases (principally N_2, O_2, Ar and water vapour).

In order to differentiate between cryogenic gas and the same gas occurring naturally in the atmosphere, an oxygen probe is inserted into the gas stream from which the cryogen composition can be inferred. By additionally monitoring gas temperatures, the

Fig. 15.4. Liquid nitrogen freezer at infeed ends of tunnel and extract duct. Product: breaded plaice. (Source: Integral Cryogenic Equipment Ltd)

volume concentration can be transformed to mass and, ultimately, into weight of cryogen emerging from the tunnel per unit time relative to product weight. An example of a nitrogen freezer at the in-feed ends of the tunnel and the extract system, is shown in Figure 15.4.

Factors Causing High Cryogen Consumption

For any application it is possible to calculate the cryogen to frozen product ratio. The leasing charge of a tunnel is generally a minor component in the cost of freezing cryogenically and, in some instances, the tunnel may be offered at a pepper-corn rent in order to tie an end user to a particular gas company. In specifying tunnel size, dimensions of a tunnel for a given belt area (i.e. nominal capacity) are important in terms of running costs, as can be judged by the cryogen:product ratios (LNR) shown in Table 15.1.

Table 15.1. Effect of tunnel dimensions on freezing efficiency

Tunnel dimensions (belt width x length)	Actual LNR	Cost/lb p	Daily consumption lb	Capital cost £
20in x 70ft	1.44	4.44	17,752	70,000
24in x 60ft	1.40	4.32	17,322	63,300
30in x 45ft	1.36	4.20	16,795	55,500
36in x 40ft	1.35	4.16	16,657	55,080
48in x 30ft	1.33	4.10	16,398	51,780
(oversize tunnel)				
48in x 40ft	1.41	4.35	17,436	63,300
(Extended dwell-time of 30 min and reduced throughput to 790 lb/h)				
48in x 30ft	1.47	4.53	9,289	51,780

In-going temp 15°C. Frozen temp -18°C. Throughput 1540 lb/h.
Belt loading @ 80%. Duration of shift @ 8 h. All-in price of LN_2: £68/t
Product: asparagus with calculated LN requirement of 1.13 lb/lb product

The analysis in Table 15.1 clearly indicates that for a given belt area:

i a short squat tunnel is more economic in freezing unit weight of product - in fact, over an 8 h period a long narrow tunnel (20 in belt width) uses in excess of half a tonne more LN_2 than a short wide (48 in belt width) one, equivalent to, £42 daily excess use for the gas price indicated;

ii the capital cost (or leasing charge) of a long tunnel is greater than that for a short squat tunnel (ca 35%).

Therefore, the end user should insist on obtaining the shortest tunnel consistent with the space allocation for the freezing operation.

Other factors capable of contributing to high cryogen costs are listed:

i Low belt coverage;

ii Over-freezing of product;

iii For a cooked or blanched product, input temperature may be unnecessarily high;

iv Too much air may be drawn into the tunnel to produce excessive internal frosting of turbulence fans and, thus, reduce heat transfer. In addition, some of the cold energy of the cryogen gets diverted to freezing atmospheric moisture;

v Blocked spray-header nozzles giving rise to lateral imbalance of heat-transfer;

vi Cut-out of a fan motor;

vii Excessively-fast withdrawal of spent gases manifested by low extract temperatures;

viii Too low a set temperature;

ix Loss of insulation due to water penetration;

x Forgetting to drain away accumulated water from inaccessible - regions of a freezer following a hose-down;

xi Short production runs;

xii Several or long stand-by modes when product is not being fed into a tunnel;

xiii Variable flow rates of product through a tunnel due to problems upstream or downstream of the freezer.

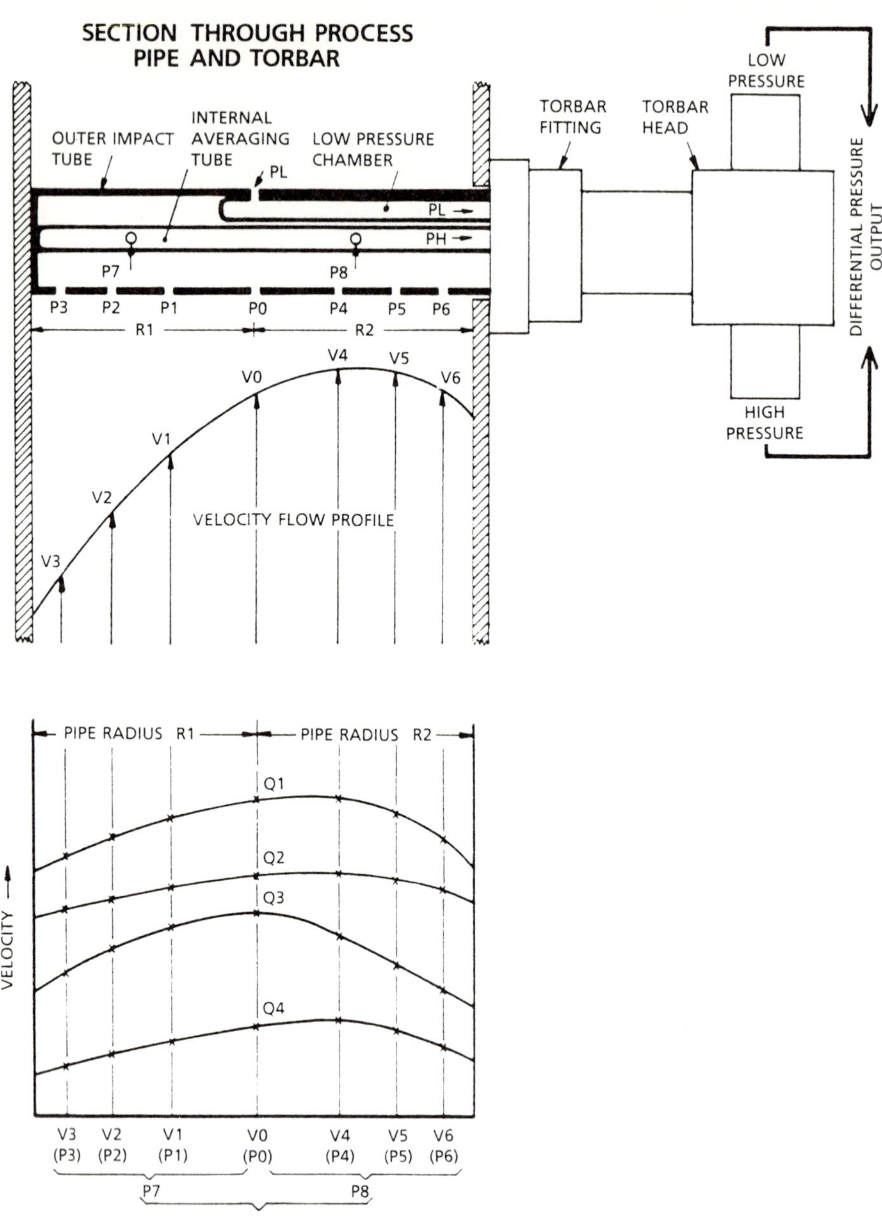

Fig. 15.5. Averaging pitot tube. (Source: PSM Flow Ltd)

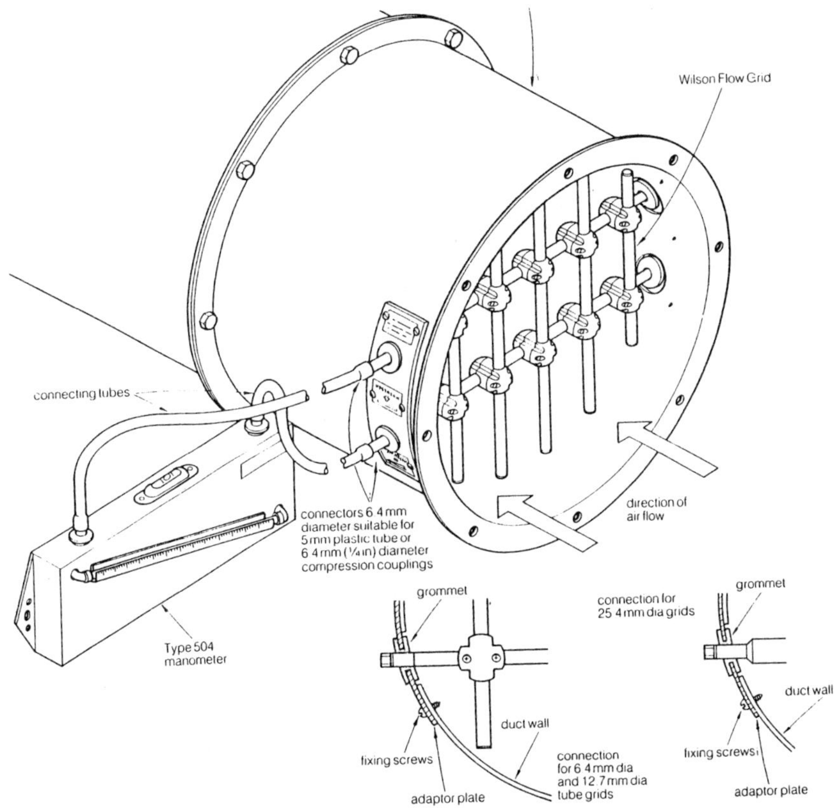

Fig. 15.6. Typical installation in circular duct. (Source: Airflow Developments Ltd)

Awareness of high gas useage is an essential step in eliminating wasteful use and facilitates diagnosis of the many possible causes of waste.

Selection of Instrumentation

Fortunately, the technology of precisely monitoring in-line gas flow and composition are well developed and practiced in other fields so that we are really tackling a technology-transfer opportunity.

Oxygen probes having a high level of accuracy and sensitivity (<0.1% discrimination) are available from a number of suppliers, as are temperature probes.

Flow metering is a complex subject in its own right. In selecting the type of flow meter used a number of criteria require to be satisfied. The probe must be robust and, preferably, have no moving parts. It must not be capable of being damaged from a piece of ice released from partial thawing upstream of it. It must be inexpensive, to justify its investment and generate the sought-after cost savings, and be accurate and easily installed into the pipeline of the extract ducting. It must be insensitive to vibrations set up by the extract fan. As the data will be processed in conjunction with other parameters (temperature, oxygen concentration, product throughput, etc) a 4 ma to 20 ma signal must be provided.

The type of flow meter that most closely satifies these criteria, is that based on differential pressure measurement. To comply with BS 1042/2A the flow grid, or averaging pitot tube (Figure 15.5), would need to be calibrated in-situ to achieve + or -1% error. Excessive swirl in the duct can cause errors, but this is eliminated by the addition of an upstream flow-straightening device. The following Table indicates expected accuracies with system obstructions using the Wilson Grid shown in Figure 15.6.

Table 15.2. Expected accuracies with system obstructions for differential flow-metering (Wilson Grid)

Obstruction	Straight upstream length to maintain + or -2% accuracy	Straight upstream length to maintain + or - 10% accuracy
Right angle bend	10D	3D
Radius bend r = 1D or less	5D	2D
Opposed blade damper	5D	2D
30° bend	3D	1D
Tapered contraction	2D	1D
Sudden contraction	5D	3D

The flow straightening distances can be reduced without sacrificing accuracy by resorting to in-situ calibration. The pressure-differential probe can be used for air velocities between 1.5 and 30 m/s and remains unaffected by vibration and movement. In the Wilson Flow Grid the unit consists of a row of tubes with closed ends, parallel to each other and forming an open fence across the duct at right angles to the flow axis. Some of the tubes are perforated with small holes facing upstream which sense total pressure, whilst the other tubes have holes on the downstream side to sense throat substatic pressure (The static pressure, if required, would be sensed at right angles to the axis of gas flow). The spacing of the holes conforms approximately with the log Tchebycheff distribution for direct summing and averaging. The upstream and downstream tubes are connected to separate manifolds which provide two average pressure signals and the difference between the manifolds constitutes the output signal.

The Wilson grid is the application of Bernoulli's Theorem which states that the total energy contained in a moving fluid is constant. Neglecting friction losses, this energy is the algebraic sum of two pressures: the velocity (or dynamic) pressure and the static (or duct) pressure. These two components can interchange freely depending on the geometry of the duct and any obstructions being present. The flow grid creates a local reduction in the free area of the duct and the increase in velocity between the tubes results in a corresponding reduction in static pressure which is measured at the downstream holes. The forward-facing holes sample the duct total (velocity) pressure and the difference provides the differential output. It is found experimentally that the ratio of the differential pressure to velocity pressure remains fairly constant over the useable velocity range. The velocity (or dynamic) and sub-static pressure over the area of the duct are measured from averaged values to provide a differential output pressure of between 2.4 and 3 times the duct velocity pressure, depending on size and type of unit.

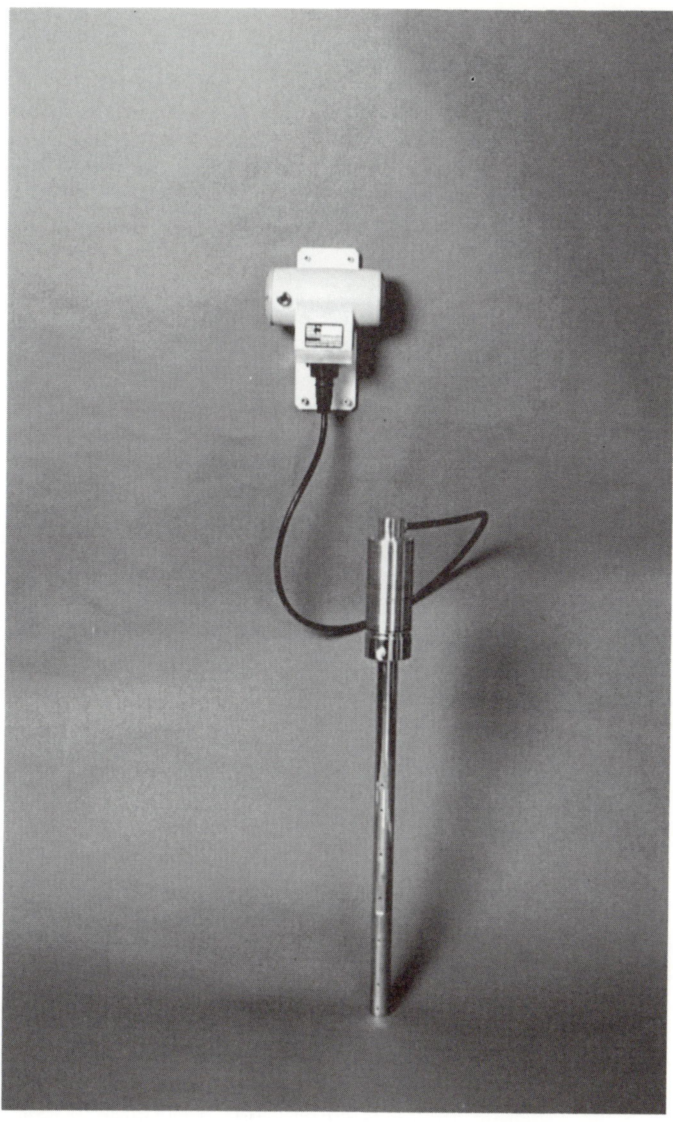

Fig. 15.7. Torbar. Averaging Pitot probe with integral dp sensor and remote mounted electronic transmitter with 4-20 mA output. (Source: PSM Flow Ltd)

The volume flow rate Q for the Wilson grid can be calculated from the emirical relationship:

$$Q = K.(dp)^{0.5} \tag{1}$$

where dp is the differential pressure and Q is the volume flow rate given by:

$$Q_v = A.V \tag{2}$$

V is the mean duct velocity.

An example of a differential pressure flow sensor is the Torbar (see Fig 15.7, a diagrammatic representation of which is shown in Fig. 15.5) which is manufactured to fit exactly across the diameter of the extract pipe and consists of four basic components:

i An outer impact tube;
ii The internal averaging tube;
iii The low pressure chamber;
iv The head.

The outer impact tube has a number of pressure-sensitive holes facing upstream which are positioned at equal annular points in accordance with a log linear distribution. The total pressures developed at each sensing hole by the impact of the flowing medium are then averaged by the internal averaging tube to give the high-pressure component of the differential pressure output. The low pressure component is generated from a single sensing hole located on the downstream side of the impact tube. In both types of dp meters (Wilson and Torbar) turn-down ratios of better than ten are readily achieved without sacrificing accuracy. The Torbar is claimed to have an accuracy of 99% + or -1% with a repeatability of 0.1%. Because of the averaging characteristics upstream and downstream, short straight pipe runs are possible without significantly sacrificing accuracy.

The volume flow rate may be expressed in the form:

$$Q_v = 67.295 \times k.A\{dp.P_a S(T+273)\}^{0.5} \tag{3}$$

And the mass flow rate for liquids, gases or steam as:

$$Q_m = 4.635 \times D^{0.5}.k.A \times dp^{0.5} \tag{4}$$

where Q_V = Flow (m^3/h) at normal temperature and pressure of °C, 1 bar;

Q_m = Flow (Kg/h)
k = A constant characteristic of the dimensions of the pitot tube used for sensing the static and mean dynamic flow pressures, and is supplied by the manufacturer.
D = Density at actual conditions (Kg/m^3) (Base density of air at °C is 1.293 Kg/m^3 and at 15.5°C 1.223 Kg/m^3)
A = Pipe internal x-section area (cm^2)
T = Actual temperature (°C).
P_a = Actual pressure absolute (Bar A).
dp = Differential pressure (mbar).
S = specific gravity at 0°C, 1 bar (air = 1)

The total cost of a computerised monitoring system for cryogen, which would include a dual bar graph digital display and alarm levels, is estimated to be less than £5000, the exact cost depending on the size of the extract duct. In practice there would be a reference oxygen probe sampling the atmosphere which would also provide a continual safety check on oxygen levels in the event that leakage of cryogen into the factory atmosphere is occurring. Alarm systems would be installed to shut off plant if the oxygen level falls below what is regarded as a safe value, a safety practice that should not be disregarded when working with LN_2.

Most LN_2 tunnels do not operate at their most economic level for the multitude of reasons some of which were highlighted. A saving in gas consumption of 15% to 30% can give rise to savings of several thousands of pounds of gas - e.g. £15 000 to £30 000 on an annual gas bill of £100 000. Even with a modest saving, we are therefore talking of repayment on a CMS of well under a year. The greater the yearly gas requirement, the shorter is the pay-back period.

Theory of Determining Cryogen Consumption Rate

The principle of operation of the CMS is to measure the total volume flow rate in the extract duct containing atmospheric gases and the spent cryogen. The method is equally applicable to carbon dioxide or nitrogen simply because the cryogen gas is inferred from the oxygen level in the duct. However, for convenience, let us assume a nitrogen freezing application. Assuming the dry atmosphere contains 20.8% v/v oxygen, then an indicated oxygen level in the duct of this amount means that no cryogen gas is passing from the tunnel into the extract duct. Any negative departure in oxygen level from 20.8%, means that cryogenic gas is diluting the atmospheric oxygen being drawn through the duct. Once the temperature is known the volume throughput of the cryogen can be converted to mass of gas and ultimately compared against the rate at which product passes into the tunnel to give the cryogen:product ratio: this is a direct measure of the efficiency of the tunnel to cool the product to the set temperature from a given input temperature. The rate of throughput of cryogen can be shown to be given by:

$$LN_2 = \frac{28 \times 273 \times Q_T \{Oa - Od\}}{1000 \times T_K \times 22.4 \times Oa} \, Kg/min \qquad (5)$$

where Q_T is the measured flow rate in litres/min at temp T_K (Kelvin), Oa is the oxygen concentration in the atmosphere, and Od is the oxygen concentration in the extract duct.

Equation 5 may be simplified to:

$$LN_2 = \frac{0.341 \times Q_T \{Oa - Od\}}{T_K \times Oa} \, Kg/min \qquad (6)$$

If the throughput of product is continually monitored - e.g. any moulded product such as a hamburger - then this parameter can be fed into the computerised calculations to provide continuous monitoring of LNR. With other products the throughput rate may need to be fed into the computer manually to arrive at the LNR.

Effect of Atmospheric Humidity

When atmospheric gases get drawn into the tunnel (by accident rather than by design) and also up into the extract duct (by design to prevent seizure of the extract fan), the moisture naturally present in the atmosphere precipitates out as a fog. The implications of this are that the measured flow will register a slightly low value. In addition, the oxygen probe will indicate a low value, the combined effect being that the inferred value of cryogen consumption will be understated by as much as 10%. For this reason the gases sampled, both from the atmosphere and the extract duct, are dried before they are passed over the oxygen sensor. At 10°C the amount of water in an atmosphere having 40% relative humidity is 0.3% w/w relative to the dry weight of air; at 16°C this increases to 0.4%. Therefore the error on flow measurement of the chilled air containing the humidity as a fog suspension, can either be disregarded, or, alternatively a small software correction can be applied.

The CMS can also be calibrated to yield an empirical constant embracing this correction (the constant 0.341 in Equation 6, was derived) by calibrating in-situ, thereby increasing the accuracy of the inferred value of cryogen throughput and exploiting the high precision capable with the sensor probes. To obtain maximum accuracy, the oxygen probes are calibrated with pure dry air. Temperature effects on the oxygen sensors are avoided by linking the duct oxygen probe and that sampling atmospheric air, back-to-back with both probes sampling air that has been cooled inside the extract duct to bring the moisture levels to a common low value. In practice, this is achieved by drawing the atmospheric air through a copper tube that passes along the inside of the duct, filtering out precipitated water droplets and then passing the dried air across the oxygen probe. Alternatively, because the amount of gas required for measuring the oxygen level is so minute, the gases could be dried by the passing them over with a desiccant such as silica gel or calcium chloride.

Conclusion

In summary, the case for a cryogenic freezer operator to be able to monitor continually the LN_2 consumption is based on the following:

1 It provides a management control facility for cost-effective consumption of cryogen by ensuring that the freezer line operates optimally at all times, without hidden losses adding to the gas bill. Excessive LN_2 useages of 30% +, are not unknown;

2 It provides a basis for the construction of a number of control loops for automating optimal tunnel performance;

3 By logging and accumulating total LN_2 consumption whenever the tunnel is in use, it provides an independent check on how much gas's delivered into the customer's tank, rather than how much left the site of manufacture of the LN_2.

Subject Index